T0344649

George Westinghouse

George Westinghouse:

Gentle Genius

Quentin R. Skrabec, Jr.

Algora Publishing
New York

ISBN-13: 978-0-87586-506-5 (trade paper)
ISBN-13: 978-0-87586-507-2 (hard cover)
ISBN-13: 978-0-87586-508-9 (ebook)

Library of Congress Cataloging-in-Publication Data —

Skrabec, Quentin R.
George Westinghouse : gentle genius / Quentin R. Skrabec, Jr.
p. cm.
Includes bibliographical references and index.
ISBN-13: 978-0-87586-506-5 (trade paper : alk. paper)
ISBN-13: 978-0-87586-507-2 (hard cover : alk. paper)
ISBN-13: 978-0-87586-508-9 (ebook : alk. paper) 1. Westinghouse, George, 1846-1914. 2. Westinghouse Electric & Manufacturing Company—History. 3. Inventors—United States—Biography. 4. Electric engineering—History. I. Title.

T40.W4S57 2007
620.0092—dc22
[B]

2006025961

Front Cover:

Printed in the United States

To my grandfather, Louis Frank Skrabec, and Our Lady of Montserrat

Acknowledgements

Special thanks to the George Westinghouse Museum and Ed Reis for the use of the archives and conversations. Critical to my research were the excellent digital archives of the Heinz Museum of Western Pennsylvania.

Table of Contents

Introduction

I first heard the details of George Westinghouse's amazing engineering career as a freshman in engineering at the University of Michigan. That may not seem too unusual, except I spent most of my early life going to school in the shadow of Westinghouse's Union Switch and Signal plant in Swissvale, Pennsylvania (a borough of Pittsburgh) and a few miles from the site of the Westinghouse mansion in Homewood. I had crossed the Westinghouse Bridge in the family car often on the way to the then new Monroeville Mall, without a thought of who or what it was named after. In fact, while I knew many people who worked at Westinghouse Air Brake and Westinghouse Electric, I didn't realize that "Switch and Signal" right down the street was a Westinghouse company.

Pittsburgh was still a Carnegie town in the 1960s. The great mills, which still turned day to night with smoke and colored the night sky with a fiery aurora borealis, drew the youthful minds toward Carnegie. Stories of Carnegie and working at his great mills were part of the local urban legends. The Carnegie name was everywhere, on plaques, signs, libraries, museums, awards, stamped on millions of books, and part of the local lore. Carnegie was Pittsburgh. Every child soon had to ask his parents what or who is Carnegie? That question would trigger many stories. Of course, every child's rite of passage included a first trip to the Carnegie Museum in Oakland and its great dinosaurs, one of them even named after Carnegie. The Carnegie name was immediately associated with dinosaurs, millionaires, the steel plants that surrounded you, and libraries. Westinghouse was merely a company name or in those days an uninteresting household appliance. In the summer I visited the Carnegie museum and library almost weekly played in Frick Park, and gazed at many great bronze statues around Schenley Park and Oakland. I loved the history of Pittsburgh and knew all the local stops of George Washington and General Braddock. I went often to Philips Conservatory, the Allegheny Observatory, and all the local sites; I loved to ride the streetcars instead of the bus through the many old boroughs of Pittsburgh; went often with my grandfather to downtown Pittsburgh to hang out at old Fort Pitt; spent hours watching the many trains and the fiery mills operate. I had also spent many hours at the Buhl Planetarium, where as a youth I received an award from Werner Von Braun for building a small steam engine science project (my grandfather did most of the work). But I never visited the Westinghouse Memorial in Schenley Park, the site of the old Westinghouse mansion in Homewood, or the town of Wilmerding, until three years ago.

The Westinghouse Memorial is not easy to get to. Since 1930, by-passes have been added and traffic in the Schenley Park area has increased. I visited on a Friday when employees of Carnegie Mellon University took up the only parking available, but it was worth the effort. I had wanted to write about Westinghouse since college, but there were an unusual amount of warnings. Francis Leupp, Westinghouse's first biographer, in 1918 warned that Westinghouse "left behind him no dairies, no files of personal correspondence, and scarcely any other sources of supply on which the biographer . . . depends." His other biographer, Henry Prout, had the help of a committee in 1921, but still complained He left no written record except in the files of his numerous companies. He wrote almost no private letters. He kept no journals or even notebooks. He made but a few addresses and wrote few papers." Even more recently, in 2003, Jill Jonnes, author of *Empires of Light*, warned again of the lack of biographical material. Recently, however, historical material from the period has been digitalized, particularly at Pittsburgh's Heinz Historical Center, Cornell University, and Rutgers, and that offered a means to penetrate deeper. There had also been a renaissance in the writings and life of Nikola Tesla, which opened new material on Westinghouse. I had crossed Westinghouse many times in these massive digital files as I was writing my earlier book, *The Boys of Braddock*. The Westinghouse Museum had been formed in Wilmerding, and had some of the old Westinghouse Electric archives. Of course, the "war of the currents" had been well researched by a number of authors, which at least served as a base to build on. My own interest was to study the nature of the man, the social pioneer, the engineer, and the manager. There is no better place to start that quest than the Westinghouse Memorial.

The Westinghouse Memorial was dedicated on October 6, 1930. The memorial was the result of voluntary donations of over sixty thousand employees of the many companies he founded. It was meant to be a reflection of the man they knew. Today it is a quiet spot surrounded by major roads for that section of Pittsburgh, but any visitor can feel the spirit of a great man. It is hard to believe it was once the site of the Pittsburgh Zoo. As you approach the Memorial, you first encounter a pond that forces you to look first from a distance of 20 yards or so. What appears is a youthful bronze statue looking a three wall panels. At first look, I must admit I didn't get it. I assumed the boy was a youthful George Westinghouse, which it is not. It is a youth looking at Westinghouse's accomplishments for inspiration, and that is the real point of the memorial. The youth looks on to three panels of granite and bronze. The dedication reads:

> George Westinghouse accomplished much of first importance to mankind through his ingenuity, persistence, courage, integrity, and leadership. By the invention of the air brake and automatic signaling devices he led the world in the development of applications for the promotion of speed, safety, and economy of transportation. By his early vision of the value of the alternating current electric system he brought about a revolution in the transmission of electric power. His achievements were great, his energy and enthusiasm boundless, and his character beyond reproach; a shining mark for the encouragement of American youth.

Certainly, no one could put such statements of character on any of Westinghouse's contemporary industrialists. It would have been unlikely that Carnegie's employees would have ever erected a memorial or make such a statement of greatness, yet Carnegie remains the Jupiter of the industrial gods. While Westinghouse may have a lesser place in the Pantheon of Pittsburgh's Industrialists, his place in the hearts of his employees was unequaled. This is the first message you take away from this Schenley Park memorial. It may explain why the real enduring memorial to Westinghouse in the Pittsburgh area was the happiness of those who worked for his companies. This imprint that he left on his organizations had to be part of the story.

The main wall had the famous picture of Westinghouse at his beloved drafting board, with an engineer and a worker on each side. The two sidewalls each had three of Westinghouse's accomplishments. No real surprises here. Its early morning and the temperature is already in the 80s, and it's a workday. I find myself alone with plenty of time to think about the project, and how to find a man that's been dead for almost a hundred years. I had already read the three main biographies of Westinghouse, but wanted more of the man that was reflected in his employees. Francis Leupp wrote his biography four years after Westinghouse's death in 1918. Leupp did an excellent job on the character of Westinghouse but let the technology for some "future pen." In 1921 Henry Prout addressed the engineering feats from an engineer's perspective. Both men struggled with presentation. Westinghouse's life like the memorial was a story of major projects. His life was, however, a multitasked effort that meant much overlapping, besides the written record of Westinghouse's life appears as a series of project. A purely chronological approach could be just as confusing with the overlapping. Both Leupp and Prout opted to use a project approach, but since I wanted to emphasize the man, a more integrated story approach would be needed. The third biographer, I. E. Levine, had boiled Westinghouse's life down to an excellent story for youthful readers. The memorial offered more of the spiritual outline I was looking for. Here at the memorial we see the life of Westinghouse as seen through that of a schoolboy. The memorial was straightforward as well. Writers and biographers of Westinghouse can quickly be absorbed by the nature and difficulties of the electrical engineering, and then lose sight of the man. I realized this difficulty as an engineer and writer. I had studied (or really struggled) through electrical engineering at Michigan, and in my first assignment at National Steel, which was to improve a series of electrical steels. Yet, this was the world of George Westinghouse and it cannot be avoided.

The main panel also has an engineer, which Westinghouse made a true professional. The engineer appears to be holding a degree. Westinghouse through not a graduate engineer became a major source of employment for graduate engineers. Westinghouse promoted the development of electrical engineering programs at Ohio State and the University of Michigan. He would hire the first woman electrical engineer. On the other side of Westinghouse is a worker (mechanic) with hammer in hand. The model for the worker was Thomas Campbell, who was at the memorial for the 90th anniversary of Westinghouse's birth in 1936. Campbell in 1936 was known as "Westinghouse's oldest living employee" having started with Westinghouse Air Brake in 1869. In 1936, he reported he had been receiving a monthly pension for six years, and he was living "comfortably with my son." Mr. Campbell was living in a Wilmerding home, the company had helped him purchase and had a company life insurance plan. At the bottom of the main panel is a marble relief of a steam locomotive, which had launched Westinghouse's career.

The side panels consisted of his six major accomplishments, but they seem secondary to the inspirational theme of the memorial for America's youth. The left panel first illustrated "Railway Electrification," probably the lesser of the six, but in 1930, it still offered great potential. "Steam-Electric Power," again, a technology that has morphed, but fundamental to the evolution of the industry. The steam turbine is still basic but the coal as a fuel is at least temporarily out. Next is "Hydro-Electric Power," which Westinghouse pioneered at Niagara Falls, and remains key to the electric power industry today. The right panel highlights the "Chicago World's Fair," which is an enduring part of Westinghouse's legacy. It was at Chicago in 1893 that AC current demonstrated its utility. Next is the "Air Brake," for which he is probably best known. Finally, there is the brass plaque for "Automatic Signaling Devices," another of his long remembered contribution. I loved the memorial, and make a pilgrimage to it with every trip to my hometown. This day I will follow the old 67-streetcar line, I had ridden so often as a kid. The great streetcar lines of Pittsburgh lasted into the 1980s, but are gone now.

My next stop was a few miles away at the location of Westinghouse's mansion known as Solitude. Today it is a park with nothing of the mansion left except a stable house, which is used for offices. It is here that Westinghouse entertained presidents, princes, and famous scientists. Westinghouse's neighborhood has lost a lot of the magic of its former residents, such as Carnegie, the Mellon family, and Henry Clay Frick. The memories of the Gilded Age can still be seen in some of the old mansions. Frick's mansion of Clayton has been restored, and it has the old spirit of the neighborhood. The stories of Westinghouse wiring Frick's mansion, Teddy Roosevelt's visit to the neighborhood, Frick and the Mellon art collecting, the weekend poker games of the neighborhood, and views of the Gilded Age can be found here. Behind the mansion is one of the entrances to Pittsburgh's Frick Park, where I spent so much of my youth, and you can still get a sense of the rural setting where Westinghouse built Solitude. Frick Park still has a wonderful mix of trees and wildlife going back to an earlier era. It is difficult to find Westinghouse the man here, except in some of the great stories available at Clayton.

A mile or so away, at the other end of Frick Park is Swissvale, home to Westinghouse's Switch and Signal. A block away from the old plant is my old grade school of St. Anselm. Swissvale is still dominated by its railroad tracks, which Westinghouse used to commute to his many east side plants, as well as start his journey to New York or Washington. Swissvale is a mix of new and old, but most of the old charm is gone. In Westinghouse's day it was a rural outpost of Pittsburgh, with some worker housing, he had started. Many of the buildings still reflect the dirt and grim of industrial Pittsburgh. Though Swissvale is well past its post-war industrial boom days, many of the streets have remnants of these better times. A few more miles and you arrive at Turtle Creek Valley and the towns of Pitcairn, Wilmerding, and East Pittsburgh. Here are the ancestral homes of Westinghouse Air Brake at Wilmerding, and Westinghouse Electric at East Pittsburgh. It is here that George Westinghouse is most missed. The close of so many plants in this valley and surrounding burghs is heart breaking for any American engineer. Only the "Castle," now the Westinghouse Museum can revive you and bring memories of the great times of these valleys, but it is here in the archives that you can find the spirit of Westinghouse the man. It is here that a small glimmer of hope can be found in the legacy of Westinghouse. That hope will always be that some future Westinghouse will find their inspiration here. Finally, two hundred some miles away at Arlington Cemetery you get the last glance of Westinghouse the man in the simple tombstone. He is remembered here only as a naval officer. A future archeologist would find no evidence of his industrial achievements noted at his Schenley Park Memorial, yet it is here that you get an insight into the man.

Today, I'm at the Westinghouse Memorial almost two years later, and thousands of hours of research later. Now there seems to be something missing. His employee relations, of course, but maybe this is represented by the fact this was an employee-donated memorial. Still, I would have included a plaque on his employee relations, which pioneered pensions, life insurance benefits, health care, good housing, and educational benefits. These are what his legacy in the workplace has been. Maybe the town of Wilmerding is in itself that memorial. Beside the on looking school boy, I would add the politician and executive of today. Westinghouse represents the model for striving in a global market. Westinghouse was a free trader, but one who has created more jobs for Pittsburgh then even Carnegie. He had plants the world over in cheaper labor markets, but his American operations grew at a faster rate. The secrets were invention, innovation, employee loyalty, respect for intellectual rights, and aggressive research and development. Maybe Westinghouse's resentment of industrial financiers, foresaw their role today as corporate managers. And just maybe we are missing the great industrial lions of the Gilded Age in our struggle with globalization, these lions being the operating and engineering managers and employees, which have the innovative keys for our future.

Chapter 1. The Panic of 1907

"Well, gentlemen, this only compels me to do something else."

It started on a smoky and cold March morning in the Pittsburgh suburb of Homewood. The Westinghouse mansion porch was thick with the mill dust of the last 12 hours, and it would be a real challenge for the house staff to clean things up. Westinghouse had been up late working on the drawings of a new turbine engine to generate electricity in the den of his suburban mansion known as "Solitude." This night, even his usual companion, his butler, had gone to bed leaving him alone with his wife's dog into the early morning. He would, however, rise routinely at six-thirty after five to six hours of sleep. The morning air had a strong smell of sulfur from the steel mills, which reflected prosperity in the region. Westinghouse Electric had been working at full capacity, and was about to record quarterly earnings, as were most of America's great corporations. The morning newspaper carried the news of significant spring flooding along the Monongahela and Allegheny rivers as Westinghouse readied to follow his daily routine of meeting the first shift at the plant. George Westinghouse's neighbor, Henry Frick had just received an urgent letter to come New York for an emergency meeting of the J. P. Morgan banking firm. Letters of concern continued to flow all day long (mail was delivered seven times a day in this influential Pittsburgh suburb) as he prepared to take the train to New York. J. P. Morgan, America's premier banker, had just left for an art-collecting trip to Europe. In the days prior to the Federal Reserve, the House of Morgan controlled America's financial well-being. Morgan, a few days earlier, had been in Washington to assure the then President Teddy Roosevelt of the strength of the American economy. Henry Frick had been a friend of Roosevelt, having had a visit by Roosevelt to Frick's Pittsburgh estate, even though, many believed Roosevelt to be the root cause of all the urgent calls to Frick on this day. Another attendee of this emergency meeting in New York was Jacob Schiff, major banker to the Union Pacific Railroad. Schiff had predicted financial problems a few months earlier, "such a panic . . . as will make all previous panics looks like child's play."

The stock market was falling rapidly that morning, and the Dow Jones average would lose 25% by month's end. The situation had reached crisis stage with little hope of stabilization. While the New York exchanges struggled, regional markets, such as Pittsburgh closed for lack of liquidity. Interest rates were spiraling out of control, as stock prices plummeted. It was becoming clear that this panic was more serious then the "Rich Man's Panic" of 1903. It would also be much different than the great "Panic of 1893." This time the currency problem had international roots. Currency problems in Europe had already started a wave of gold hoarding. Even world markets, such as Japan, had a currency shortage. At the time only a few Americans realized what was taking place in the world's financial centers. This day, however, the word was slowly getting out. The New York bankers were starting to worry. Only a week ago Westinghouse had returned from New York, where he had discussed a major bond offering planned for the summer. Westinghouse was also preparing his stockholder speech for March 31, 1907, where he planned to announce a dividend of over two million dollars. To most of the nation, there was no hint of any banking problem on the horizon. These crises tended to start in small New York circles, moving in months to the average workingman in the industrial centers. On this March day, the first shock waves went out, and the sulfuric smell of prosperity would soon end in the Pittsburgh air.

A few weeks later, Westinghouse was boarding his special Pullman car on the Pennsylvania Railroad for the ten-hour trip to New York. His Pullman car consisted of a dining room, kitchen, sleeping quarters and office, including a drafting desk and engineering office. It was named "Glen Eyre" and was always ready for a business trip. His journey would start from a tunnel at his mansion that went directly to the train station. George had a drawing board in his Pullman for he still found relaxation in drawing and outlining mechanical devices. Westinghouse was a large man, giving the appearance of a walrus with his Victorian mustache and long coat; the chairs in his car were specially designed chairs. He had designed his train car to be a home, and indeed, he sent many hours in that car. Westinghouse was a modern commuter, taking the train every day from his Pittsburgh home to his plants around the area and his downtown office. On winter weekends, he connected to a Pennsylvania express train and went to another home in Washington D.C. On spring, summer and autumn weekends, his destination was his home in Lenox, Massachusetts. Most Monday mornings, his train took him to his New York office at 120 Broadway. He went abroad twice a year, always visiting his office at the Westinghouse Building in London.

The only comfort Westinghouse could find this day was the hiss of his own air brake system of the train. Certainly, that reminded him of a simpler time where he could get lost in his love of engineering. Westinghouse was making his first of many long trips to New York. This meeting would be with several New York bankers, including Morgan's firm, to look at a bond offering for Westinghouse Electric. One of the meetings would take place in Morgan's home library, where most the nation's financial decisions were made. He knew he faced an uphill struggle as the panic was now taking root in the late spring of 1907. The Pittsburgh banks were feeling the crisis and were reluctant to loan more to Westinghouse, who already owed them $9 million. Even the powerful Mellon family of Pittsburgh turned down his request for more credit. He had some friends in Mellon Bank, but the first shock wave had dried up all local capital available.

Mellon Bank's failure to help is surrounded by rumors to this day. The Pittsburgh bankers cited Westinghouse's liberal employment policies, which were far out of line with practices of the time. The Westinghouse pension plans, disability insurance, housing aid, etc., caused concern both in Pittsburgh and New York. There was a rumor that Pittsburgh bankers wanted Westinghouse to fail because of bad blood going back to the 1890s.[1] In 1893 Westinghouse had refused a Pittsburgh demand to place one of their associates on the board of directors. They had insulted Westinghouse as well, saying that he lacked spending control, and they even implied he was a poor businessman. Westinghouse lost his temper, reminiscent of his childhood, and stormed out of that Duquesne Club meeting. The New York bankers were more than happy to bind Westinghouse to them with debt at that time. Andrew Mellon, the dean of Pittsburgh bankers, never forgot how Westinghouse rejected his request and went to New York.

It is interesting that Henry Clay Frick was a Morgan man and was in the "Morgan library meetings" that sealed Westinghouse's loss. Frick was one of the Mellon family's closest friends as well; he was a partner with Mellon in Pittsburgh National Bank, and he was also a neighbor and at least a casual friend of Westinghouse. Westinghouse was one of a few friends including Andrew Mellon and Philander Knox (a lawyer for Carnegie Steel and secretary of State under President Taft) to be part of Frick's Pittsburgh card games. Westinghouse had also helped Frick convert his nearby mansion of Clayton to electricity. In the bitter personal and legal feud between Carnegie and Frick in the 1890s, Westinghouse had tried unsuccessfully to mediate a peace. Frick was clearly in a position to save Westinghouse, but did not. Frick was always a businessman and banker first. He, like Morgan, believed in a corporate model of management while Westinghouse represented the paternal system. Even Carnegie's patriarchal system was preferable to the al-

1 Haniel Long, *Pittsburgh Memoranda* (Breton Books: Pittsburgh, 1939), 42

most socialistic approach of Westinghouse. Westinghouse could only hope that the other New York banks would help as they had done in the 1890s, but he was pessimistic. His strong opposition to the banking and corporate trusts of the day made him a liability and a target of big business interests. Westinghouse left no records, and biographers of Frick, Westinghouse, Morgan, and Mellon all ignore Frick's role, but Frick could have made the difference — that much is clear.

Westinghouse had few friends among New York bankers, even though fifteen years earlier they had helped him out of a similar financial crisis. That crisis had opened the door to more control of his company by bankers. Still, it had been worth it to win the lighting contract for the Chicago Columbian Exposition of 1893. The event led to the use of AC current throughout the country versus DC current. Westinghouse had triumphed over the Morgan-backed Edison Electric, but he had made many enemies in the process. Westinghouse had further provoked the anger of the New York bankers by his refusal to involve Westinghouse Electric in trusts and combines. Westinghouse opposed any type of trust as an obstruction of free competition. Even worse, Westinghouse was a supporter of Teddy Roosevelt's trust-busting policies. Westinghouse had even become outspoken on the topic, further enraging the New York financial barons. J. P. Morgan and other New York bankers had been unsuccessful at pulling Westinghouse into industrial combines and alliances. Westinghouse in particular had rebuffed Morgan in 1890 in an offer to control competition in the electrical industry by price and output agreements.[2] Westinghouse wanted independence. What Westinghouse was opposed to would be revealed in the 1912 congressional Pujo Committee hearings. The findings of the Pujo Committee were unbelievable. The officers of the five largest New York banks held 341 directorships in 112 major corporations. Morgan-related officers were on 72 company boards. Westinghouse Electric had been the notable exception to the New York bankers' control of American industry.

Westinghouse was an extremely profitable company; in fact, in 1907 it would report record profits. The problem was that Westinghouse had borrowed heavily to finance new projects and development work. It was this debt and high interest payments that concerned the bankers. Still, in May of 1907 Westinghouse had some leverage. First quarter profit had been $2.8 million dollars, but because of his dividend rate used to maintain investment, almost $2.5 million went to cover dividends. The high dividend rate allowed Westinghouse to raise capital quickly for his big projects. He had depended on bank credit as well, and had many corporate bond offerings. He was also considering another stock offering to raise capital, but the word was out in financial circles that Westinghouse could not be expected to hold a 10 percent dividend rate. In general, Westinghouse had little interest in this financial stuff. J.P. Morgan represented a new style of banker. Morgan was

2 Jean Strouse, *Morgan: American Financier* (Perennial: New York, 2005)

interested in building companies that could generate profits and interest for his banks. Morgan had successfully done that in 1900 with the formation of United States Steel and the formation of General Electric. Westinghouse wanted to invent and build things. The banking barons of New York had little interest in Westinghouse's engineering projects, their interests were in profits. Morgan's modus operandi was to take control of successful companies with strong operations, then put in a financial organization that would send the profits to the banks.

J. P. Morgan found it difficult to understand men like Westinghouse. Physically Morgan, being over six foot tall, could stand up to a determined Westinghouse (who was often intimidating). Morgan even had a fascination with electricity and cutting edge technology. Morgan had been an early backer of Thomas Edison. His Madison Avenue home was the first New York home to use Edison's incandescent bulb system in 1880. Morgan was, however, never a satisfied customer, investor, or business partner of Edison. And in a slow but methodical process, Morgan took over Edison's General Electric by 1892, leaving Edison as a figurehead only. Morgan couldn't understand Edison and Westinghouse's paternal management style. Morgan also lacked trust in Westinghouse because of Westinghouse's application of "costly" Christian principles in the workplace. Morgan tended to separate his Christianity from his businesses. Westinghouse's implementation of the half-day Saturday holiday seemed frivolous and costly to Morgan. His use of corporate money to fund employee pension plans represented a dangerous precedent. He had even built the true "company town" to house his employees, known as Wilmerding, near Braddock, Pennsylvania. Similarly, Westinghouse's establishment of a "Relief Department" to help disabled workers struck Morgan as out of place. After all, if Westinghouse Electric was struggling, how could such huge sums of money be designated for employee programs? No other major corporation had any of these unnecessary expenses. Another enigma was Westinghouse's belief in competition and avoidance of trusts, which could have made his companies more profitable. The idea that some social view could get into the way of business was incomprehensible to Morgan. He could understand the love of engineering; he had seen that in Thomas Edison. But socialism and capitalism seemed incompatible to him. To Morgan, this was not the mark of a successful capitalist, but of a socialist or communal manufacturer, such as Robert Owen in Scotland.

Westinghouse's spring trip to the banking empire was at least temporarily successful, but he had increased his debt by another fifteen million with a major storm gathering. Morgan seemed to have been convinced by fellow Pittsburgher, Henry Clay Frick, to support a Westinghouse corporate bond offering in the summer. Westinghouse felt confident that he had averted a crisis. Of course, Morgan realized that more Westinghouse debt meant more control by the bankers. With the crisis averted, Westinghouse turned back to engineering in the summer of 1907.

The project on his drawing tables was a steam turbine to power ships. He had already harnessed the power of Niagara Falls with turbines, but now he recalled his very first invention, that of a rotary steam engine. In 1900, Westinghouse pioneered the first turbine driven electrical generator for Hartford Electric in Connecticut. The dream of a rotary steam turbine to drive a ship went back to his days as a ship engineer on a Union blockage ship during the Civil War. He had studied the reciprocating steam engines that were still used in 1907 to turn propellers. The reciprocating engine had to turn back-and-forth motion into rotary motion at a great loss of power. A steam turbine could use rotary power to turn the propeller shaft directly. Seems simple enough, in fact, Herodotus developed the first steam turbine in 300 BC. As a boy one of George's first engineering projects was the building of a windmill, which is a true rotary engine. Even I had with my grandfather built one as a sixth grader using a tin can. There was one problem. A steam turbine moves at extremely high speed, while the ship's propeller moves at a very slow speed. The problem was therefore to merge these speed differences. This would require a type of engine transmission to reduce rotary speed to a more useful speed/power combination.

Westinghouse had recently purchased a patent of an imperfect steam turbine by Charles Parsons of London. The Parsons steam rotary turbine had been used to drive electrical generators for lighting ships in1895. Parsons's turbine was being used on the marine ship called *Turbinia.* Still, the Parsons turbine operated at too high speed to drive large marine propellers. It was a typical strategy of Westinghouse to purchase patents and improve on them. Using this approach he had purchased the Tesla patents of electrical generation, and developed an alternating current electrical distribution system in the 1890s, as well as an electric motor. Westinghouse dreamed now of a turbine engine that could develop more power with less fuel. Furthermore, a steam turbine would reduce space needed and allow longer cruising times without repairs and refueling. Such a turbine would be a highly efficient user of energy. Another part of Westinghouse's strategy was to bring in expertise. In this case, he hired the retired Chief Engineer of the Navy, Rear Admiral George Melville, and marine engineer John Macalpine. He already employed Benjamin Lamme, who had built the first Westinghouse turbo alternator. He had the people and the dream in place in 1907, and that is where he wanted to be, mentally. He could always find relief in his drawing board. Financial concerns were usually left to his vice-president, H. Herr, and his financial secretary Walter Uptegraff.

Summer brought new challenges to Westinghouse and his companies. The panic was spreading in banking circles and the stock market was showing signs of a bigger problem. Westinghouse's bond offering was failing as the banks tightened and the stock market fell. August was particularly problematic with the crash of the Egyptian Stock Exchange at the start of the month. Bond offerings for the cities of Boston and New York were also selling poorly. Metropolitan Traction Company, an iron manufacturing concern, went bankrupt in early August as well. The problem was spreading internationally, as well, with the Bank of England overextended trying to help the Egyptian and Japanese exchanges. This was becoming the first "international" banking crisis for the United States. Teddy Roosevelt's trust busting Attorney General had won a big case against Standard Oil, which had further frightened the bankers. On August 10, 1907, the American market crashed. J. P. Morgan rushed to return from Europe at the end of the month as the whole of American banking was near collapse. Since there was no Federal Reserve to move in, J. P. Morgan would have to supply some form of leadership. Morgan telegraphed his agents, such as Frick, to start developing strategic plans. Panics were opportunities for men like Morgan and Frick. Frick had built his coal empire with Pittsburgh bankers, the Mellons, in the Panic of 1873.

September brought some relief in the stock market, but the banking crisis was actually worsening. Westinghouse Company had good sales and cash flow, and it was preparing to pay a dividend in October. In fact, Westinghouse Electric had been paying a 10% dividend rate since 1903. The problem was its $43 million in bonds and debt. In particular, the company had fifteen million dollars of floating debt coming due. Even with record profits, Westinghouse could not cover such huge debt if called in by the banks. The bankers were trying to protect their depositors and had no credit to extend. In addition, Westinghouse's refusal to join the banking trusts of the day found him outside the big banking circles. For J. P. Morgan and the New York banks, this was an opportunity to crush one of the industrialists that had remained outside the trust. The Pittsburgh bankers lacked the resources to help. If Westinghouse came under the control of the bankers, only Edward Libbey's glass empire of Toledo would be free of trust control.

In October, the stock market slumped again. October 13 was considered by Morgan a lucky day, and he was clearly preparing for the opportunities that might arise from a financial panic. Morgan may have consulted with the Archbishop of Canterbury, which he often did, for "mystical" signs. On October 14, 1907, the stock of United Copper fell from $62 a share to $15. The loss forced owner F.A. Heinze to pull money from his Butte, Montana savings and loan, which affected a chain of banks. Nine banks pooled together to aid in the United Copper problems. Then a run started on Knickerbocker Trust Company because its president, Charles Barney, had a close relationship with F.A. Heinze. Depositors withdrew over $8 million in less than four hours. On Friday, October 18, the run reached a point at which all backers pulled out, thereby assuring its failure. The stock market saw a number of companies go under that Friday as well as United Copper Company. The weekend added to the tension as stories of bank failures hit the nation's press. Westinghouse was at his New York office that Friday as the market signaled a crash. He studied his books with his financial secretary Walter Uptegraff. The debt was over $40 million for Westinghouse Electric alone. The immediate danger was that the banks would call in short-term loans, which was typical of financial panics before the Federal Reserve. Four million dollars in cash would be needed by midweek. Thursday had been payday in the Turtle Creek Valley, but Westinghouse would have to reduce the payroll Monday to get all the cash possible. Saturday morning, October 19, Westinghouse left for Pittsburgh, still confidant that quick temporary moves would work.

Monday morning October 21, the city of Pittsburgh braced for the arrival of the New York financial tidal wave. The word had hit the city that payrolls around town would be affected, including those of the booming mills of Carnegie. Ernest Heinrichs, Westinghouse's public relations man, described the mood in the Westinghouse Building on October 21; "an atmosphere of ominous oppression pervaded the offices. Conversations were carried on in whispers. Everybody seemed to have a feeling of fearful expectancy, as if some catastrophe was about to descend upon the place. Although nobody appeared to have any idea what was going to happen." The same feeling existed at Carnegie Steel Headquarters. Cash calls would quickly become cancelled orders and layoffs for Pittsburgh's industries. Carnegie Steel had been taking order cancellations all morning. Tuesday morning the news came by telegraph that Knickerbocker had failed, which meant the banks would be calling in loans.

Interestingly, J. P. Morgan's refusal to aid Knickerbocker on October 22 caused the panic to proceed, but now Morgan could direct its path. Morgan chose to organize aid for Trust Company of America. Westinghouse was again the target of press articles claiming poor management. That day he went to lunch at the Duquesne Club with his Public Relations man Ernst Heinrichs. Heinrichs had been hired back in the fall of 1889 to help fight Edison Electric's attacks on the danger of AC current. This fight was tougher because Heinrich, a former industrial reporter for the *Pittsburgh Chronicle Telegram*, could find no help in the local papers. With Morgan's takeover of Carnegie Steel in 1901, Pittsburgh followed the lead of the Morgan-controlled New York press. Besides, Westinghouse was an outsider to the Scotch-Irish bankers who controlled the city's financiers, such as Andrew Mellon and Henry Clay Frick. They met briefly at the Duquesne Club with Judge Reed, who they hoped might act as a bridge. Reed had been Carnegie's chief lawyer and had worked with Morgan to negotiate the formation of United States Steel. Westinghouse's old friend and now Mellon Bank lawyer, Judge Reed, would try heroically to pull together Pittsburgh bankers to help.

Judge Reed was one of the most respected and well-placed businessmen in Pittsburgh. He was a director of United States Steel, president of Bessemer & Lake Erie Railroad, director of Farmers National Bank, and president of Reliance Insurance Company. He had the business ties to bring both the Pittsburgh and New York bankers together. Reed could bring Mellon into a Westinghouse deal, but a green light would be needed from Morgan's New York Library. Success was almost in hand when, on October 22, Knickerbocker Trust Company failed. That really was the start of the Panic of 1907.

The failure of Knickerbocker started a chain reaction that only J. P. Morgan had the power to stop. Reed was in the end a Morgan man and would always defer to Morgan's wishes. Westinghouse had taken the train to downtown Pittsburgh that morning. The headlines of the Pittsburgh paper were about a strike by women clerks in Pittsburgh department stores. The previous day had been a slow one for news across the nation, so that the same story was on the front page of the New York papers. Lillian Russell was opening her new show, *Wildfire*, at the Nixon Theater in downtown Pittsburgh. The sports pages were still talking of Honus Wagner's winning of the National League batting title. Westinghouse arrived at his office at 8 am. Later that morning at the Westinghouse Building in Pittsburgh, telegraph boys were running in and out of this financial hornet's nest. Phones were ringing and were ignored. Reporters were everywhere asking questions. Westinghouse issued a press release: "The Company is not insolvent — only hampered for the moment. It is doing more business than ever before. It will come out alright."[3] Westinghouse had bought stock personally, two days prior, only to see his investment drop 40%. By mid-afternoon, the Pittsburgh skies were as dark as night due to the smoke of the local steel mills, which were straining to meet booming demand. At 5:30 in the afternoon, Westinghouse called his Vice-President H. T. Herr and his lawyers to start drawing up receivership papers. Amazingly, Westinghouse found a few minutes to discuss plans for his new steam turbine with his chief engineer, but it was a forced effort to remain in control. Westinghouse was doing his best to keep his employees throughout the Pittsburgh area calm.

In New York on October 23, 1907, J. P. Morgan returned to his office to find Henry Clay Frick waiting for him. The panic had hit the streets, with bank depositors pulling their money, since in those days there was no protection of bank savings. Some creative individuals earned as much as $10 a day to wait in bank line as depositors tried to salvage other parts of their lives. The draw down on bank reverses caused banks to call in loans from big and small companies. In Boston, retailers posted "panic prices" to help generate cash. Morgan held day and night meetings in his library as he moved to take control of the panic. Morgan even forced the banks to issue scrip in lieu of cash to keep the banking system floating. Lacking any government regulation, the public looked to Morgan as their savior. It was an image promoted by the *New York Times*, as Morgan controlled most of their debt. Even Teddy Roosevelt had no choice but to ally with Morgan. Morgan's European connections promoted the image as well. Europe was also near collapse, and renowned banker Lord Rothschild called Morgan, the world's greatest financier.

Morgan had no interest in saving Westinghouse. Besides, J. P. Morgan was aliened from his biggest rival, General Electric. This was the overall strategic approach of Morgan, which awaited only the opportunity to activate.

3 Francis E. Leupp, *George Westinghouse: His Life and Achievements* (Boston: Little, Brown, and Company, 1918), 210

Furthermore, Westinghouse had soundly defeated Morgan, Edison, and General Electric in the "war of the currents."

Morgan was now ready to close the trap on the only major industrialist that had ever successfully rebuffed him. The meeting with Frick and others on the situation went all through the night, which was typical of Morgan's "Library meetings." The Morgan meeting moved to stabilize the nation's banks, while also picking some industrial plums as part of the harvest. Frick and the United States Steel Company moved to purchase Tennessee Coal and Iron. Westinghouse was left to another consortium of New York bankers to devour. Westinghouse in Pittsburgh knew the banks could not be stopped. The Pittsburgh banks were limited to help since all looked to New York for saving cash, and New York banks were failing every hour. Andrew Mellon, the lead Pittsburgh banker, remembered being rebuffed by Westinghouse in the Panic of 1893, and held back the Pittsburgh bankers as well. In fact, this day the Pittsburgh Stock Exchange closed for three months due to a lack of capital liquidity. Still, as always Westinghouse had employees and stockholders on his mind. This would affect them far more than himself. Westinghouse Air Brake and Union Switch & Signal Company were safe. However, Westinghouse Electric, Westinghouse Machine, and Nernst Lamp Company would be lost. This night he would pass on hosting dinner at his mansion. Most nights, it was common to have a dozen or more quests, often including visiting engineers, ambassadors, railroad men, and famous scientists. These dinners had become famous throughout Pittsburgh. This time, George wanted to be alone. October 23, 1907 would be the beginning of the end for one of America's greatest and kindest industrialists.

Westinghouse was just one of the casualties of the credit market panic. Today's control of such markets by the Federal Reserve would have spared a profitable company like Westinghouse Electric. But, in 1907, there was only J. P. Morgan and other the leading bankers who pooled $25 million from the government, $10 million from John Rockefeller, and $25 million from New York banks to stabilize the market. Morgan took charge (and took advantage) of the financial crisis. By the end of the month, he was being hailed as the nation's savior. An ode to honor Morgan ran in the *New York Times* on October 27. Morgan, however, was a major benefactor of the crisis. He had knocked out most of his competition. United States Steel (a Morgan-backed company) took over its major competitor. Westinghouse Electric, General Electric's major competitor (GE was also a Morgan company), was taken over by Morgan. On November 13, Charles Barney, former president of failed Knickerbocker Trust, killed himself with a gunshot. Allegedly Barney was despondent over Morgan's refusal to meet him. The real measure of Westinghouse's character was the offers of his employees and small investors to help him in this crisis. It was the type of loyalty that would never be known by Morgan, Frick, and Carnegie. Yet, the big eastern papers hailed Morgan as the nation's savior.

Morgan's altruism was not praised in many of the industrial towns across America. Morgan had successfully taken over huge chunks of America's industrial might and thousands of jobs. J. P. Morgan was to knock out another potential competitor in Nikola Tesla. Morgan a few months prior had formed the "Alaska syndicate" to control copper mining and supplies. He was even working on cornering the rubber market needed for electrical wire insulation. Tesla had just announced a major research effort into wireless electrical transmission and was waiting substantial financing from Mercantile Bank. The Mercantile Bank was one of the first Morgan that Morgan had allowed to go under. With these victories Morgan controlled steel, banking, copper, mining, railroading, and the electrical power. He even had his hands in oil and rubber. Morgan's triumph during the two weeks of the panic became the target of congressional investigations in 1909 and 1911. Records showed that there was a long array of visitors to Morgan's library nerve center, including United States Steel's financial committee. Judge Gary and Henry Frick had worked long hours with Morgan's accountants to buy out Tennessee Coal, Iron, and Steel Company. TC&I was a major steel competitor of United States Steel. In 1907, TC&I was one of the stocks in the Dow Jones average and a very visible national company. Prior to 1907, Morgan and USS had feared antitrust suits if they moved to take over. The market crash also allowed TC&I to be purchased at bargain prices. Morgan's role in the Westinghouse failure was less defined and had less scrutiny.

Even President Teddy Roosevelt kept his distance from Morgan. Roosevelt and the Republicans were the great trustbusters, which Morgan championed. It was a love–hate relationship for Roosevelt, who publicly lambasted Morgan at every opportunity, but needed him in financial crises. The government had the money to support individual banks, but these panics tended to create a chain reaction. This would leave the government endlessly chasing the problem. In 1907, the country needed a national bank or Federal Reserve, as Alexander Hamilton had suggested in the 1700s. Lacking such a national bank, the government was dependent on people like Morgan to pool banking resources in panics. This pooling in effect created a national bank, but it required leadership that only Morgan could supply. Roosevelt sent his Treasury Secretary, George Cortelyou, to represent him at the Morgan library meetings in the last weeks of October. Yet Roosevelt would avoid giving Morgan any public credit.

In fact, most populists and Republicans of the time suspected that Morgan had caused the panic. Wisconsin's populist Senator Robert La Follette stated in 1907 that Morgan and his associates "deliberately brought on the late panic, to serve their own ends." John Moody charged that Morgan and associates stopped the panic by "taking a few dollars out of one pocket and putting millions into another." Upton Sinclair used the panic as a basis for his novel, *The Money Changers*, to illustrate Morgan's ruthlessness. In the novel a Morgan-like character orchestrates a financial crisis for private gain while destroying ordinary people across the nation.[4] The Senate committee investigating the panic for years concluded Morgan had taken advantage of the situation. Teddy Roosevelt testified in 1909, no longer as President, that he had given tacit approval to Morgan and Frick to proceed for the good of the country. The testimony took the energy out of the investigation. Even the view in Pittsburgh was mixed because of the gain for Morgan backed United States Steel. The press played it by sympathy for Westinghouse the human being while placing the financial failure squarely on Westinghouse's shoulders.

In the end, however, the loss of Westinghouse's companies was decided in Morgan's library, not from his poor management. Three of Westinghouse's companies, the electric company, the machine company, and the investment company went into receivership. Morgan had his revenge, but it was control that really drove him. Westinghouse had repeatedly refused to enter into a trust with General Electric. By Morgan's rules he had no choice but to crush Westinghouse. Westinghouse Air Brake, Union Switch and Signal, and Westinghouse Canada were all debt free and had plenty of cash to ride out the panic. Westinghouse had found himself with much debt coming due at the very moment of the panic, allowing the bankers to control destiny during the crisis. Companies in much worse financial condition survived by being favored by the bankers. Personally, Westinghouse Electric had become his baby; it was the company he was most attached too. The loss, therefore, was very personal. Many of Westinghouse's associates appeared to have left him in his time of need. He had belonged to the right church (Presbyterian), right clubs (Duquesne Club), and the right political party (Republican). He was part of the Pittsburgh industrialists, but he could never be a part of the inner circle of Carnegie, Mellon, Frick, Pitcairn, Oliver, and other key money men. In particular, the Scotch-Irish controlled the banks and there was a lot of reciprocity among them. Henry Clay Frick had to be the most disappointing to Westinghouse, since Frick had ties to both Mellon and Morgan.

In late October of 1907, Westinghouse was on a train to New York once again, this time to discuss receivership with the bankers. He rode separately in his car from the Pittsburgh bankers who were invited to the meeting. He was never comfortable traveling, preferring an evening at home, and this day was even more trying. The train passed the Air Brake Works and the town of Wilmerding, which he had built. Wilmerding looked from the train as a great industrial castle surrounded by well-planned rows of houses. This was Westinghouse's model town that rivaled the famous industrial community of New Lanark, Scotland. It had only been a few weeks ago that he had signed the order for thousands of turkeys to be distributed in Wilmerding for Thanksgiving. The tradition went back to 1870, when in the early years he had invited all his workers to Thanksgiving dinner at a Pittsburgh hotel. As he opened the morning paper, the train rolled by his huge complex

4 Jean Strouse, *Morgan: American Financier* (New York: Harper Perennial, 2000), p 589

of Westinghouse Electric and Machine companies in Wilmerding and East Pittsburgh. The front page of the paper discussed Westinghouse's poor financial management. Westinghouse had always enjoyed the support of the local press, and it hurt deeply to have it turn on him. He noted to his associate, "I suppose all those great works built themselves."

The fact was Morgan that controlled and was favored by the press nationally, which assured many more difficult mornings for George in the months to come. Papers that had once hailed his kindness and character now portrayed him as a fool at every chance. It was tough medicine, but it was part of the image building of Morgan, the corporate white knight. The paper, on his arrival in New York, gave Westinghouse even less comfort; the Morgan controlled New York Times, Sunday edition of October 27 hailed Morgan with the ode:

A millionaire is wicked, quite
His doom should quick be knelled;
He should not be allowed to grow
If grown he should be felled,
But when a citys' bonds fall flat,
And no one cares for them
Who is the man who saves the day?
It's J.P.M.
When banks and trusts go crushing down
From credit's sullied name
When Speechifying Greatness adds
More fuel to the flame,
When Titan Strength is needed sore
Black ruin's tide to stem
Who is the man who does the job?
It's J.P.M.

In a few years J.P.M. would be answering to Congress about whether he had "saved" the country or created the problem in the first place.

The hardest and most unfair criticism of Westinghouse was made by a jealous Morgan associate, Andrew Carnegie. Carnegie came to the United States from his home in Dunfermline, Scotland, during the end of the Panic. Carnegie's remaining investments and friends were protected from the Panic by the hand of Morgan. Feeling a bit superior and lacking details, he was his usual gregarious self on the deck of the ship. Carnegie quotes made the newspapers and the following quote was typical, "I feel very, very sorry that George Westinghouse should have trouble and I want to see him out of the woods. Fine fellow, George, George is — splendid fellow. And a great genius. But he is a poor businessman. A genius and a businessman are seldom found in one individual. Now Westinghouse is of too much value to the world, in originating ideas and developing them, to have one whit of energy wasted in business work and worries. You see, all of his business activity would never get him individually a noticeable success, whereas his genius, at play, would keep home an outstanding figure in the world. He should have a good business man, so that he never would have to bother about business details."[5] Carnegie was just plain wrong. Westinghouse had done things Carnegie had never done as a businessman. Westinghouse started over thirty companies from scratch and built them into multi-million dollar international companies. Westinghouse started companies in twenty different countries.

In New York, Westinghouse would sit down with the conservative lawyer Robert Mather, who had been appointed chairman. Westinghouse was to remain on as company president. The new board represented bankers from New York, Pittsburgh, Chicago, and Boston. Mather wanted Westinghouse out and at first suggested a six-month vacation while they reorganized the company. Westinghouse was not interested, in fact, he was anxious to continue the company's large program of research and experimentation. He realized these bankers would have none of that. Only a decade prior, J. P. Morgan had wrestled the "Electric Wizard," Thomas Edison out of his company — Edison Electric. These New York bankers wanted manufacturing cash flow without the outflow of research. Shareholder equity was Morgan's measure of success. Research is unnecessary if you have no competition.

He returned to Pittsburgh to talk things over with his wife, Marguerite. Westinghouse had been the lion of industry for years. Marguerite had been in the shadow of her husband. Now he needed her support at the age of sixty, as his dreams seemed to bending. He could have no better friend and wife than Marguerite. Their marriage had been full and loving through the best and worst of times. She sold all her gifts of stock to assure that her husband would know no embarrassment in this short-term money crisis. She suggested a ride in their "horseless carriage," which had been a gift from his French Westinghouse Works. He was hurt, depressed, and disillusioned. He had started to believe the editorials, which blamed him for the company's demise. That very day, the editorial had been titled "The Dynamo is stilled." Marguerite listened and encouraged George as he struggled. The next two years would be a personal hell. Westinghouse battled against a boardroom bully, usually coming out on the short end. Westinghouse was suffering from a type of depression known well to men of achievement. Without Marguerite, he would have given in to it, but her support kept him in the fight. Westinghouse stayed on as company president, but he was restricted from any financial decisions. It was difficult to see the innovation that had been the touchstone of Westinghouse Electric shelved for short-term profits. He fought each battle, but this was not the Westinghouse of old. The bankers now wanted Westinghouse out; they had hoped that he would leave as a

5 Greater Pittsburgh Chamber of Commerce, printed by Robert Forsythe Company, 1928

gentlemen. Morgan modeled his takeovers after Napoleon, his hero, allowing local control but requiring strategic loyalty. Morganization was usually a long process. Organizations and cultures survived as long as money flowed to the stockholders.

The re-organization of Westinghouse Electric under the auspices of the bankers went smoothly, since the issue was debt management, not profitability. This is what distinguishes the Westinghouse re-organization from most of those the reader of today may be familiar with. In fact, even within a few months after receivership, Westinghouse was on the rise. Even the re-organization depended on Westinghouse himself. Westinghouse invested 1.5 million of his own money. His employees as always showed their loyalty with 5000 of them contributing $611,000 for a new stock issue. Some smaller banks and brokerage houses also courageously helped out as a personal favor to Westinghouse. The bankers realized that the Westinghouse name was key to the company success, so that even their victory was not complete. The company retained its name, a concession that the bankers had not even given Thomas Edison, when they took down Edison Electric in 1891. Westinghouse remained President of the company, but he no longer had full control. The big bankers picked the new executive committee, and named Robert Mather, chairman of the new Board. Westinghouse never fully accepted these bankers in charge of "his" company. T. Hart Given, President of Farmers Deposit National Bank, and one of the receivers of Westinghouse Electric told the story of Westinghouse's cold approach to them. Many times Westinghouse and the Pittsburgh receivership representatives, such as T. Hart Given and William McConway, were called to New York. Westinghouse would travel on the same train with his private car, but allowed these bankers to sit in regular coach. The bankers restricted Westinghouse to sales and operations, restricting product development and research dollars. This was what he had always feared, and had been the root of his financial strategy to keep the bankers out.

Though Westinghouse was too big a name for the bankers to publicly force out, he felt compelled to hold on to what little he could — not for himself, for he was crushed mentally, for he was crushed mentally, but because he felt obligated to his employees. Morganization meant the Westinghouse social advances would be repelled. They had hoped he would have resigned, but he stayed. His pride was wounded and his need for control assaulted, but he stayed. He had seen the ouster of Edison and the cold corporate evolution of General Electric. Westinghouse surely wanted to resign, but he felt an obligation to the stockholders and employees. Personally, it was extremely hard on him, but he fought through his personal depression. The bankers surrounded and isolated him, and made his life hell. They vetoed the money for many of his projects. The strategy was typical of Morgan and the big bankers of New York. It had taken almost ten years to route the Carnegie men out of United States Steel after the buyout of Carnegie Steel in 1900.

In the summer of 1911, the directors forced a proxy fight for the company president. The papers were following this ordeal as Westinghouse lost to Edwin Atkins. Thousands of workers in Pittsburgh were disappointed at the loss of their beloved boss. Marguerite and George soon came to the same conclusion; it was time to break from Westinghouse Electric. He was too old to battle twelve bankers everyday. He needed to accept the fact the company was lost. Now, he would take that vacation to his summer home in Lenox, Massachusetts. Lenox was where Westinghouse had built his "recreation house." He would go there every summer with Marguerite, whose health demanded she leave the hot sulfuric air of Pittsburgh. At Lenox, George Westinghouse passed the time hunting, fishing, gardening, and bowling on his built in lanes. He would never be the lion of old, but neither could he retire from his passion for work and innovation. He had made the decision to work through his depression because he could not bring himself to accept the defeat of retirement. The defeat cut deep, however, and Westinghouse never fully mentally recovered. His resentment over the loss pushed him into a mix of depression and a power drive. He was driven to prove himself, being highly affected by the comments of the press about poor management.

Westinghouse was a man of achievement. Achievement represented a passion and an internal core of the man. Losing Westinghouse Electric amounted to only a small dent in his personal fortune, but took its toll mentally. With all his kindness and humility, Westinghouse was a deeply proud man, passionate about winning and achieving. Control of his companies met the freedom to pursue his projects as a hobbyist. Westinghouse Electric could no longer research areas of his interest. It was now more than a setback; it was the start of a mental prison for him. This type of pride is common among achievers, and what distinguished Westinghouse was his commitment to play by the rules. He had resisted the bankers demand that he form "alliances" of companies to control the market. In a time when human labor was cheap, Westinghouse treated his workers as he would want to be treated. His concern was real and went deeper than just motivating his workers. This is what differentiates him from other Victorian paternal capitalists such as Andrew Carnegie. But, like Carnegie, the bankers would never eliminate their mark on the companies. Westinghouse's vision prevailed at Westinghouse Electric, and that was the mark of Westinghouse's true character.

Chapter 2. In Search of Character

Most biographies focus on a great inventor's creative youth and the development of his mechanical skills, but the real gift of Westinghouse's upbringing was character development. Character defined Westinghouse's successes, failures, and his legacy. Westinghouse's personality and kindness resulted in loyal and highly motivated employees, which in turn built him an industrial empire. The inventor Nikola Tesla, who had been taken in by many of the bankers of the time as well as by Westinghouse's competitors, said, "He is one of those few men who conscientiously respect intellectual property, and who acquire their right to use the inventions of others by fair and equitable means." This practice allowed Westinghouse to pool the creativity of many inventors into mega-inventions. Samuel Gompers, founder of the American Federation of Labor, said, "If all employers of men treated their employees with the same consideration he does, the AFL would go out of existence." At Westinghouse's funeral, eight of his original employees were pallbearers. Former employees donated thousands of dollars for his memorial in Pittsburgh. He lost Westinghouse Electric due to his ethical stand against the trusts and bankers of the day. Westinghouse's belief in fair play in business would be outstanding in any age. One observer said, "It was, perhaps, his upbringing that set his lifestyle; he grew up in an environment of work, thrift, and responsibility."[6]

---

6 "We shall not look upon this again," *News & Views*, George Westinghouse Museum Foundation, Edition 2005-3

The Westinghouse family was German, of Saxon-related stock, with the original name being "Westinghausen" or "Westinghas." One branch of the family migrated to England from Leipzig, Germany, and ultimately this branch came to America. George Westinghouse Sr. was a second-generation New Englander born in North Pownal, Vermont in the New York/Massachusetts corner of the state (25 miles from Troy, New York), placing the family roots there prior to the American Revolution. He was a large man over six feet and described as "big boned." His wife, Emmeline Vedder, of Dutch-English stock, was also from North Pownal. George Westinghouse Sr. followed the family tradition of farming. Soon after their marriage in 1831, George Sr., and his wife moved to the Ohio frontier in search of new opportunities. He purchased farmland on the banks of the Cuyahoga River near Cleveland. Their stay was short, as they found the Snow Belt winters and extremely humid summers hard to bear. They moved east in1836 to Minaville, New York, and returned to farming. Minaville is about 20 miles west of Schenectady.

It was here that a neighbor purchased one of the early threshing machines, which would initiate an interest in machinery. George, like most New England bred farmers, was a self-trained carpenter and mechanic. Fascinated by the threshing machine, George helped his neighbor repair and improve its operation. These adaptations developed in him a passion for the design and manufacture of farming machines. Minaville lacked the necessary waterpower; he required opening his own manufacturing firm. Deciding to move forward with his dream, George Sr. purchased land in Central Bridge at the juncture of Cobleskill Creek and the Schoharie River about 20 southwest of Schenectady. He started as a farmer with an ancillary business in repair of factory agricultural equipment. Within a year, he was manufacturing threshing machines and seed scrapers. He continued to farm, hiring farmhands as he spent more time manufacturing. He had several patents and inventions that propelled him in the early 1840s. George Sr. would ultimately have seven patents to his name. Westinghouse & Co. pioneered the use of detailed engineering drawings and modeling in design.

When George Jr. was born on October 6, 1846, the Westinghouse family already had three daughters and three sons. Westinghouse & Co. was established at the time of his birth. George Jr., however, offered a new challenge to the family. George Jr.'s early years revealed an unusual character defect. He demanded attention and threw violent temper tantrums when not appeased. These outbursts were frequent and well known as confirmed by interviews with neighbors by Westinghouse's earliest biographer Francis Leupp.[7] He continued these outbursts in grade school, and fought other boys often. This was also a difficult period for the Westinghouse family. Two more boys, Henry and Herman were born, but died in infancy. In 1853, when George Jr. was seven, Henry Herman was born creating more competition for his parents' attention. Finally, George Jr. seemed to channel his temper and restless-

7 Francis Leupp, *George Westinghouse: His Life and Achievements* (Little, Brown, and Company: Boston, 1918)

ness into working at his father's shop. This was also surprising to his father, since none of his other boys showed any interest in mechanics. He became reclusive because the other children taunted him. A somewhat poor student, he did excel in mathematics and free hand drawing. These skills helped him take quickly to blueprint reading and mechanical drawing in his father's shop. One of the foremen, William Ratcliffe, encouraged him by building an amateur workshop. William Ratcliffe would be one of a number of people who took an interest in young Westinghouse. Ratcliffe and George Jr. built sleighs, small steam engines, miniature machines, and even tried to build violins. Ratcliffe taught George Jr. to set up a lathe, and at the age of ten young Westinghouse could machine metal parts on his own. This newly channeled energy and his older brother Albert helped to calm George Jr. Albert, in particular, helped George Jr. relate better with his classmates. Albert, the third son, had an easygoing temperament and love of books. George looked up to this older brother, who was a calming influence and acted as a model for young George.

By 1856, the growth of the family and business pressured George Westinghouse Sr. to move both. He purchased a cement mill on the south bank of the Erie Canal and moved the family to Schenectady. Ironically, Schenectady would one day become headquarters for George Jr.'s major competitor — General Electric. Westinghouse's business had also expanded into the manufacture of small steam engines.

The physical move had a beneficial impact on the life of Westinghouse Jr. The neighborhood and his Central Bridge schools had labeled him as a troublemaker. Once such a die is cast, it is hard for a youth to turn things around, even though George Jr. seemed to want to do just that. In Schenectady, he ran into a teacher-mentor Alice Gilbert, who encouraged his strengths. He first excelled in drawing and math, but with Miss Gilbert's support, he took on once difficult subjects. Achievement and small successes seemed to drive George Jr. to further achievement. He started to excel in grammar, literature, and writing. Miss Gilbert predicted greatness in his future, which even further motivated him. Westinghouse would credit her for his success throughout his life. George Jr. started to see a linkage between math and mechanics. His improved academic performance and a new set of friends seemed to have a similar improvement on his overall behavior. Westinghouse appeared to have learned to divert the energy of his anger into positive achievements, but there would be times, later in life, that that anger would break through.

The dominant influence on George's character was the "Holland Puritanism" of his mother. She gave Westinghouse a unique view of capitalism that would be the hallmark of his character and businesses. Puritan ethics were typically adapted to business. For example, look at the Puritan Catechism known as the Westminster Larger Catechism, which Westinghouse received training in. The Westminster Larger Catechism extended the following application on the 8th Commandment — Thou shall not steal:

> The duties required in the 8th Commandment include the following: maintaining truth and faithfulness and justice in contracts and commerce, between man and man; rendering to every one his due; restitution of goods unlawfully detained from the right owners thereof; . . . avoiding unnecessary lawsuits, the care to preserve and respect the property and rights of others just as we care for our own.

His mother's training stuck with him all his life. It was the reason behind his refusal to join the banking trusts of the day, as well as his fierce defense of patent and intellectual property rights. It was the reason capitalists of the time such as J. P. Morgan, Henry Clay Frick, Andrew Carnegie, and so many could not understand him. Many times Westinghouse had the opportunity to cheat on patent rights or pass on full payment, and he always chose the latter.

Westinghouse's view of capitalism defined the Protestant work ethic in this country. Westinghouse applied a type of Calvinist principle that promoted personal achievement and socialism. The view was decades later summarized when German sociologist Max Weber said, "people had a duty to work, a duty to use their wealth wisely, and a duty to live self-denying lives."[8] Westinghouse's industrial methods would augur the later theoretical writings of Max Weber.

Young George learned many values from his father as well. His father's Lutheran pragmatism blended well with the Puritan spiritual lessons of his mother. His father made him an apprentice mechanic over the protests of a reluctant Emmeline Westinghouse. George had demonstrated a preference for metal work over carpentry. George Jr. was still hardheaded and proud, which concerned his father. His father assured his supervisors that the necessary discipline could be applied to his young son. An apprentice in the 1800s was really little more than a day laborer with a potential future. George was, of course, over qualified for this entry trades position having achieved the skills of many master mechanics. Typically, the youth protested the menial work such as sweeping, but his father taught him that hierarchy of the apprentice system could not be violated. Still, George Jr. rebelled often against the system, only to find his father unrelenting. These were tough lessons for both men and reflected back to his child-like tantrums. His father continued to demand even more from his apprentice son and eventually wore him down. Moving through the trade apprentice system was depended on hard work, without any shortcuts for an owner's son. Of course, George Jr. was still a boy and returned to the shop in later evening to work on experimental models and building projects.

Just as important as his father's discipline was his example as a paternal and kindly shop manager. Of a less formal religious bent than his wife, he was every bit as convinced that hard work resulted in success. He was an avid abolitionist in a very troubled time in America, but slavery went to the heart of his belief in the dignity of man. George Sr. also respected the craft of

8 Daniel Wren, *The History of Management Thought* (John Wiley & Sons: 2005), 28

the machinists pioneering the apprentice program with other shops in the Connecticut Valley. He insisted on not working Sundays. He tried to limit long hours so that the men and he might spend more time with their families, although the shop often had to work Saturdays to keep up with demand. Certainly, young George worked Saturdays so as not to interfere with his schooling. It is rumored that he said if ever he owned his own business, he would give his employees both Saturday and Sunday off, a promise he would live up to. His father also tried to run on a cash basis, never using debt but building cash reserves to depend on in tough times. George observed his father working patiently with troubled workers as well as his father's financial generosity with these employees. It was a type of family in the workplace that made work enjoyable. Westinghouse's father's abolitionist views were reflected in his hiring and his respect for new immigrant labor including a tolerance for Roman Catholics — although in the case of Catholics, some of the kindness was mingled with the hope of converting them to a more Calvinist branch.

George Jr.'s perspective on manufacturing was affected by what some have called the Connecticut River Revolution of the first half of the 1800s. The Connecticut River Valley cuts through New Hampshire and Vermont before heading south; it really encompasses all of New England. Author Charles Morris compares it to today's Silicon Valley.[9] This was the birthplace of the American Industrial Revolution, or Machine Age. It was, however, as much a philosophy as an industrial movement. It molded Yankee self-reliance and individualism with an artisan and farming culture producing machine shops to meet the growing needs of the economy. These were farmers who became mechanics by necessity. Learning mechanics and blacksmithing to maintain their equipment would eventually lead to a new class of machine age worker, independent of the farming duties. New England rivers and streams powered these new production centers. By 1850, their mechanization of spinning and loom machinery were more productive than those of industrial Britain. The area produced some of America's greatest early engineers, such as Thomas Blanchard, Eli Whitney, Samuel Colt, Cyrus McCormick, Isaac Singer, and Decius Wadsworth. This was the ideal of Thomas Jefferson's agrarian manufacturing society. This Jeffersonian view included a self-sufficient manufacturing community. It is a view that can be seen throughout Westinghouse's career, combining ethics, self-reliance, and social conscience with diverse manufacturing skills. Westinghouse took this Jeffersonian approach further in applying it to mass production and heavy industry, where many believed the Jeffersonian philosophy could not go.

Westinghouse senior's diversity of skills and Emmeline Westinghouse's social conscience were reflected in their sons. Jay, the eldest son, never took to machining operations, but he did get an engineering degree from Rensselaer Polytechnic Institute. He developed into a solid salesman and man-

9 Charles Morris, *The Tycoons* (New York: Henry Holt and Company, 2005)

ager, and upon his father's death replaced him as the head of Westinghouse Manufacturing Company of Schenectady. John, the next oldest, had strong mechanical skills, but lacked the aptitude. John was earlier on impacted by the family religious values, and favored social work. John loved to work with the poor in the community, and best reflected his mother's beliefs. Albert lacked even a little interest in manufacturing, but he loved books and literature. George Junior's younger brother Herman was similar and idolized him. Herman invented a high-speed steam engine in 1880 to drive electric dynamos, which would be the foundation product for Westinghouse Machine Company of Pittsburgh. Herman would ultimately follow George Jr. and take over Westinghouse Air Brake when George Jr. died in 1914. In all the boys you could see the social skills and conscience of the mother and pragmatic religious views of the father. All the boys were noted in their lives for their fairness, honesty, and generosity. It was a family of character and concern for their fellow man. It is in his upbringing that you see the future inventor and socially responsible manager — George Westinghouse Jr.

During his teens, George's mechanical and drawing skills increased rapidly. He mastered all the metal working skills of his father shop including machining, forging, soldering, casting, and surface preparation. He developed a true love of mechanical drawing, which would remain with him for the rest of his life. He was able to draw steam engine parts, then model them in wood, and proceed to craft them out of metal. The combination of drawing skills and manual skills where not always common among practical mechanics of the time. Westinghouse's combined drawing and mathematical skills helped perfect his use of the graphic language of engineering. Early on George Jr. learned to start his boyhood projects with engineering drawings. Around the age of 15, George Jr. took a particular interest in steam engines and started to make drawings of a new type of steam engine, which would be rotary driven. This early research included using the Schenectady Free library and industrial pamphlets available at his father's factory. He had the scientific understanding of Thomas Edison, but could fully handcraft his science projects on his own. He was a superior mechanic to Henry Ford, as well as having an ease for book research that Ford lacked. He could read and draw blueprints, which were skills that neither Ford nor Edison ever mastered. What he shared with his inventing contemporaries was a drive to achieve.

George Westinghouse Sr. imparted an array of mechanical skills to his son, but just as important were some necessary inventing skills. Westinghouse had a number of inventions to his name and was a contributor to *Scientific American* in its early days. He had been involved in patent research as well. He was quoted in 1850 in Scientific American over a patent battle concerning water wheels. A new inventor was trying to collect royalties from machine shop owners. Westinghouse showed the history of such a design was over 30 years old. He loved to work on engineering drawings and got young George initially trained. He often took young George to visit various manufacturers. The shop had the current journals of mechanics and manufacturers available for young George to read. Of course, he taught him to repair steam engines, which he had world-class expertise in.

At the time, steam engines were complex machines known as reciprocating engines. Young Westinghouse could design, draw the parts, machine them, and assemble them by the age of 12. Today's automotive gasoline engines are also "reciprocating engines." These steam engines of Westinghouse's day used steam to move enclosed pistons up and down or back and forth through control valves. The up and down motion, or in the case of steam locomotives, back and forth motion was converted to circular or rotary motion by connecting rods and shaft arrangements. This conversion of up and down motion to rotating motion required many parts and was inefficient. The first steam engines used in threshing machines were adapted from locomotive engines in the 1840s. These were even more inefficient in that the final converted rotary motion of the steam was transferred to the thresher by a belt and pulley system, threshing being the process of removing the grain off the straw and separating the grain from the chaff. The booming railroad industry drove many advances in steam technology using a horizontal piston arrangement, and so did George Westinghouse. Through out his career, he and his brother Herman would perfect the steam engine. Westinghouse engines reflect a type of beauty seen in Victorian design approach.

George dreamed of direct steam rotary power. His library research revealed to him the original design of Hero of Alexandria known as the "Sphere of Aeolus." This Greek engine of second century B.C. was a rotary engine. It was a heated sphere of water that caused steam to come out of two jet nozzles causing the sphere to spin. Hero had also discussed a more direct use of steam to turn a wheel, but he never built one. It was this direct use of steam that fascinated Westinghouse. The work of Giovanni Branca in 1629 had also been published, which used a steam jet to impinge upon turbine blades to drive a rotating wheel. This design was similar to the water wheel that powered his father's factory. Furthermore, water turbine and turbine blade design had been advanced by French engineer Benoit Fourneyron and was being applied to power generation in the 1850s. Westinghouse ideally hoped to use steam directly to drive an enclosed turbine wheel and produce rotary motion. Young George began to sketch a rotary engine that would not require the conversion process. He started with many rough drawings proceeding over months to detailed parts drawing. Westinghouse was both an excellent artist and engineering draftsman, a skill that would be a key aid in many of his inventions. By early 1861, he built a small model of a rotary engine. Westinghouse's design was simple and direct, but was probably beyond the capability of bearing and material technology of the time. Westinghouse would return several times to this project of the years. In fact, over the years many of his company testing departments would have rotary models so that Westinghouse could spend sometime relaxing and dreaming with this boyhood toy.

Just as important as George Jr.'s study of the rotary engine was the evolving methodology of his approach. Research of existing technology would always be part of Westinghouse's design approach. He also started with freehand sketches to put his original ideas to paper. He would rework these simple drawings as the idea itself evolved. This straightforward approach came from observing machinists in his father's factory. The next step was to move to detailed engineering prints with dimensions. His companies would employ more draftsmen than any other U.S. employers. Later in life he would promote drafting as a discipline in colleges as well as grade schools. Finally, modeling of the project was used for initial testing. His friend William Ratcliffe had taught him the importance of modeling, a technique he would spend a lifetime perfecting. No one had put these techniques together in such a formal approach to design. More than anyone George Westinghouse Jr. defined the discipline of modern engineering and design.

One constant concern for the Westinghouse family during George Jr.'s youth was the growing clouds of war. The Westinghouse family had always been loyal followers of Abraham Lincoln. The family was torn by a revulsion for the killing of war and their strong abolitionist views. They believed any war would be short. Still, George Sr. held his own boys back from enlisting. Albert and George Jr. had more romantic views of war. Mrs. Westinghouse had prevented George Jr. from getting even a hunting rifle until he reached the age of 15. George Jr., however, loved hunting and had many of his boyhood trophies preserved. Still, war was a different matter. By 1862, his brother Albert joined the New York Volunteer Cavalry and John joined the Navy. John had actually been caught in the draft, and opted for an engineering officer's job in the Navy. Mr. Westinghouse had used his extensive political and business connections to obtain that commission for him.

The brothers' service started ominously when Albert was captured in the failed 1863 advance on Richmond. He was part of a prisoner exchange and then received a commission as a lieutenant in the cavalry. Tension built in New York as rumors started that Lee's army might break through in Pennsylvania. The New England recruiters were trying to build an emergency force to oppose Lee, if necessary. George Jr. was only seventeen, but the crisis changed things for both his parents and recruiters. In September 1863, George enlisted in the Army as a private. He did his early service in the Washington and Northern Virginia area, as part of the cavalry hunting for rebel raiders from the likes of Mosby's guerillas. In an effort to obtain a commission, young George Jr. worked as a recruiter back in New York. This brief assignment failed to gain him his commission and he returned to the active cavalry as a private in mid 1864. By most accounts, George Jr. never really took to life in the cavalry, where most of his time was spent feeding and caring for horses.

In December of 1863, Albert, George Jr., and John met the family at a cousin's home in New York City for a family reunion. It was a brief one for Albert, who was called to duty. Military discipline had further improved

George's social behavior and ended most of the temper problems that had so concerned his parents. The military, however, had not quenched his desire to achieve. One year later in December of 1864, news reached the Westinghouse family that Albert had been killed in a cavalry charge at McLeod's Mills, Louisiana. Albert had been George's brother and best friend. Just prior to this, George Jr. had obtained a commission in the Navy as a Third Engineer on the *Muscoota* as a result of George's drive, after having made only corporal in the cavalry, and his father's connections. The position would return George to the application of the mechanics that he loved and in the officer corps he would further develop the skills of managing men and equipment. The military routine and discipline had struck a cord in Westinghouse's rebellious heart. Later in life, it was said, "His self-control was such that his closest associates find it hard to recall any anger, but he did often disarm his antagonist by the genial smile which reflected the kindness of his heart."[10]

The *Muscoota* was a steam-powered ship, and the engine room would supply valuable mechanical experience to young Westinghouse. Westinghouse continued to build models of his rotary steam engine, and successfully applied it to small wooden boat models. His position as assistant engineer allowed free time to sketch and model engines. This talent of free hand drawing was something he had in common with many great inventors, such as Leonardo da Vinci, Thomas Edison, and Nasmyth. Westinghouse was transferred to another steam vessel the *Stars and Stripes* in early 1865. The *Stars and Stripes* participated in the Potomac Flotilla, which blockaded Southern ports. The war ended at Appomattox on April 9, 1865, and Westinghouse returned home in June, now a determined and disciplined young man. Officer training had also reinforced a deep code of ethics. However, he remained depressed over the loss of his brother Albert, and a little guilty for having missed combat. His father now encouraged him to go on to college.

George Jr. entered Union College of Schenectady in the fall of 1865. This was the college of William Seward, Lincoln's Secretary of State. Other 19th graduates and students included Chester Arthur, former U.S. President; Charles Steinmetz, famous electrical engineer; and Edward Bellamy, author of *Looking Backward*. George Jr. seems to have struggled through the fall semester in the "scientific department." He then returned to his father's shop, followed by another semester or two at the college. Little is known of his academic record, but observers again noted his love of drawing. Visual thinking is a cognitive process that Westinghouse shared with other inventors, such as Thomas Edison, Henry Ford, Henry Bessemer, Michael Owens, and Henry Ford. Where Edison and Westinghouse differed was that Edison kept all his rough and initial drawings in notebooks. Edison allowed others to take these rough sketches to blueprints. Westinghouse left no records or notebook drawing — the legacy of his genius was in the blueprints, and he drew most of them himself. Westinghouse was always more the true engi-

10 Henry G. Prout, *A Life of George Westinghouse* (Cosimo Books: New York, 2005), 301

neer than scientist. Westinghouse had the rare ability to think in various two-dimensional views as in the blueprint process itself. Most of the other inventors noted thought visually but in three-dimensional drawings.

He spent a lot of time making the final drawings and patent application for his rotary engine. He received the patent on October 31, 1865, but he never built a practical version of it. It would, however, remain a hobby like attraction for him throughout his life. Thirty years later, employees remember him playing with the model of the engine at his drafting table. One associate described his playing with the rotary engine as the equivalent of others playing golf. This simple success would, however, give him the confidence to average a patent every six weeks for the rest of his life for a total of 361. Perhaps even more important was that during his lifetime his organizations would produce over 3000 patents. His father's company, Westinghouse & Co. continued without his participation. The company continued to produce agricultural machinery such as threshers. In 1871 George Jr. took a quarter interest in the company, but his brothers Jay and John managed it. In 1883 George became a partner with his father, and the company was renamed the Westinghouse Company, but his brothers continued to manage it. George Jr., like his father would strike out on his own. George Westinghouse realized his rotary engine offered little opportunity for income. His hunger for achievement would have to be supplied by a profitable career.

Chapter 3. Railroad Inventor

> "Westinghouse once said it would take a thousand years to finish building the country west of the Mississippi River. 'But I tell you this — give the railroads a good brake, and we'll do it in twenty-five.'"
>
> — Superintendent of Chicago, Burlington, and Quincy Railroad

The railroad industry, more than any other industry, defined the 19th century explosion of technology and invention. It was the industry that birthed Edison, Carnegie, Bessemer, and Westinghouse. The railroad industry maintained almost a hundred years of exponential growth driving a world industrial boom. In 1850 there were only 9000 miles of track in America, by the end of the Civil War in 1865 there were around 35,000 miles, by 1874 there was 77,740 miles, by 1900 there was 198,964 miles, and by Westinghouse 's death there was over a quarter million miles of track. The growth of railroads caused a boom in industries, like coal, steel, iron, machining, and mining. The linking of the East and West Coast via the transcontinental connection of the Union Pacific and the Central Pacific occurred in 1869. Anything related to the railroad industry was a growth industry. By 1890, the American railroad industry's gross income exceeded one billion dollars with total capital invested over 10 billion. The weight of goods moved increased enormously, from 90 million tons in 1860 to 235 million in 1880, and 425 million in 1900.[11]

---

11 T. K. Derry and Trevor Williams, A Short History of Technology (New York: Oxford Press, 1960), 385

These are staggering numbers even by today's standards. The 19th century's rich men, such as Vanderbilt, Carnegie, and Morgan had made their money in the railroads' growth. A small invention for the railroad was a way to riches from 1870 to 1920. One growing problem was the crashes, which seemed to be a daily occurrence. These crashes were grabbing the headlines with the grim news of death and injury. Editorials across the nation were screaming for industry action on safety. The public was becoming fearful of rail travel. Westinghouse was familiar with many of the issues with the railroads as Schenectady was a key railroad center and town.

After his return to his father's firm, George Jr. became involved with customers and suppliers in the East. He commonly traveled by train to make these supplier and customer calls. He had the drive like his father to branch off on his own, and was well aware of the growth potential in railroad related products. Like most biographies of the time, Westinghouse's early biographers used a heroic approach combining individual genius with Providence. Thus the story of Westinghouse's first railroad invention is shrouded in myth and legend. Many of his early successful railroad inventions tend to be familiar stories in early biographies of Westinghouse. The reality is probably a mix of serendipity, experience, and a determined drive to invent. Westinghouse was determined to be an inventor, which combined with skills and knowledge opened many possibilities for the young man. One experience common to any railroad traveler of the time was timeless delays due to derailments. We do know that Westinghouse was a keen observer and carried a pencil and notebook. He had also decided early on that he wanted to be an inventor, and he dreamed of some breakthrough invention.

The invention that launched his industrial career was the railroad "car replacer." The portable rerailing frog or car replacer was a portable or permanent track device, which enabled derailed railroad cars to be put back on the tracks. These rerailing switches are really car replacers, using the common terminology of today. Still, crowbars and muscle were the more common method of rerailing a car in the 1860s. Westinghouse, like any traveler of the time, was familiar with both methods. Accidents and derailing were common experiences. A rail traveler could expect to experience hours of delay, with time to get out of the car to watch the muscular job to re-rail. The problem with most car replacers was that they did not hold up well under the high stresses of putting a car back on track. The poor design also gave them little advantage over muscle power in the time required to get the car back on track. The design was improved to facilitate the rerailing operation. This improvement alone would cut the time of rerailing from a couple of hours to under a half-hour. Sometimes it was referred to as a "switch" or "car replacer." The frog replacer being used to "jump," the car back on track. The device was a short piece of track that could be clamped to the existing rail section and used to bring the car back on track. The device allowed the locomotive to pull the derailed car back on the main track. Derailment was a common problem in the 1860s because of poor track bedding, increasing speed of the locomotives, and inferior tracks. Cars would jump the tracks often, which required crews to use crowbars. A simple one-car wheel jump could mean hours of delay. More serious jumps where a major reason for railroad fatalities, as another train might plow into the derailed train. The western lines were higher speed was common, often suffered from rail snaking, which caused the track to bend up catching the moving train. Wrought iron track caused most of the snaking. This observation would lead Westinghouse to use steel in his car replacer, wrought iron being the common track material until the 1880s when Bessemer steel became economical. The car replacer had probably fermented in his mind from years of living around railroads and railroaders for so many years. The success of his car replacer led him to design a more durable permanent track switch known as a frog.

The Westinghouse frog was not the first such device, but it was the first major advance in the device. The railroad frog was a piece of switching track that allowed trains to switch tracks. The frog was a permanent switch or piece of track, having a life of only a few months as speeds and car tonnage increased. The term is still used today for a segment of intersecting rail that allows one track to cross to another. These switches had to take the continual impact of trains. Wear and breakage was a constant problem, which in turn clogged rail traffic when replacement was needed. The railroads were purchasing hundreds of these each month. Westinghouse started his work on the frog sometime in 1866. His typical approach of drawing and modeling started the project. What became obvious, however, was the wear on the device. Cast iron was too brittle and wrought iron too malleable. Steel offered the ideal compromise, but the use of steel was limited by its high cost. Westinghouse hung around the railroads and had heard of steel being used on rails in England. Before he had resolved the materials problem, he applied for a patent calling it an "improved railroad switch" in the fall of 1866. He also started the search for capital to back his idea using a wooden model of his new "switch." He needed a foundry to cast this new design as well. His father's machine company did not include a foundry forcing him to search elsewhere. Interestingly, as much as his father supplied jobs, housing, and education for his boys, he never invested in any of George Jr.'s inventions. His father at least initially wanted George Jr. to stick with the farm machinery where the family had built a reputation.

At twenty years old, Westinghouse would have to bring all his sales ability to apply, to promote his new device. He was able to convince two Schenectady businessmen and friends of his father to contribute $5000 each to form a partnership. The company incorporated as Rawls, Wall, and Westinghouse in 1866, and the three partners would share the profits equally. George would remain working part-time for his father and live at home. The sales burden fell on young George to promote his new product, but first he needed a steel foundry to produce his frog. Most of steel in 1866 was produced by the crucible method, but the new Bessemer steel process was emerging as an economical alternative. The second American Bessemer steel plant had just been built a few miles away in Troy, New York. The plant was the work of another engineering genius — Alexander Holley. The plant had been partially financed by the Pennsylvania Railroad because of interest in steel railroad rails. Tests around the world had hailed the superior strength and wear resistance of steel over cast iron and wrought iron. Crucible steel took weeks of heating and processing into small batches, now the Bessemer process offered high production and lower price steel. Westinghouse realized that using steel meant a tougher, longer-lived frog. Westinghouse was able to contract this new mill at troy to built his frogs and replacers. The cast steel frog was one of the first steel castings ever made in the world, but its success required a great deal of experimentation by George Jr. Steel is cast at a much higher temperature than cast iron and has poor liquid flow characteristics, making castings extremely difficult. Westinghouse read, studied, and experimented his way to a casting method. This type of intensive study preceded all of Westinghouse's endeavors. Research included public libraries, trade journals at his father's plant, and the stories of local railroaders. In this case, his study of casting methods gave him metallurgical expertise, which he used throughout his career building some of the world's greatest foundries by the end of his career. He would also make numerous contributions to ferrous and non-ferrous metallurgy over the years.

George's first major sale of the cast steel frog was to a small independent railroad — the Chicago, Burlington, & Quincy Railroad for some trials. This contact had been secured through some local businessmen, but Westinghouse also hit the road selling each railroad. The cast steel frog also sold itself by its superior performances, with improved frog life at twenty times the cast iron product. The ability of Westinghouse to supply became a problem and he was required to add a crucible steel foundry as a supplier at Pompton, New Jersey. It appears that most of the Westinghouse frogs were crucible steel, whose smaller batches were more adaptable to castings. Bessemer steel at Troy was being used for high tonnage production of railroad rails. It may have been also that Westinghouse realized the superior quality of crucible steel over Bessemer steel. Crucible steel, while costly, is a cleaner and tougher product for difficult applications. Before Westinghouse received his first frog patent in February 12, 1867, he had applied for another on an improved version. This improved version included the mention of steel, but not the very important design feature of reversibility. The feature was really a redundant design, which provided for basically two frogs in one product. When one side wore out it could be reversed, further extending life and reducing the cost to the railroads. The design concept of redundancy so prevalent in engineering today was one that Westinghouse was a pioneered. Westinghouse would apply this redundant design concept throughout his career in very diverse areas, such as natural gas distribution and electric motor design. Westinghouse even at this young age was an amazingly skilled engineer. Superintendent Towne of the Chicago, Burlington, & Quincy Railroad was one of the first to impress the young inventor on the need for an improved braking system.

Some time in 1866, George was to meet his future wife on one of his many business trips. One story has that meeting occurring on the Hudson River Railroad on a return trip from New York City. Often the women traveled in non-smoking cars while businessmen preferred the smoking cars. This day the smoking cars were filled and Westinghouse, a non-smoker anyway, opted for a non smoking car. It was here that he met his future wife — Marguerite Erskine Walker of Roxbury, New York. Marguerite had been visiting relatives in Brooklyn and was on her way to see more relatives in Kingston. They seemed to have fallen in love that day; Marguerite was a tall, dark-haired beauty. She was outgoing and seemed to draw out George, who tended to be a bit reserved. The dating period was extremely short, and formal, with Westinghouse supplying references from friends and his pastor. Within a year they were married on August 8, 1867 in Roxbury, New York. Business required them to skip the traditional honeymoon, and they had to start their marriage living in Westinghouse father's house.

Shortly after the marriage, problems developed with Westinghouse's partners. Westinghouse was in a weak position, having no real cash position in the company. His partners tried to bully him into selling his rights to the patents. They underestimated young Westinghouse's confidence. Westinghouse's capital was in the patents, not the cash. Their tactics to force control of his frog patents was his first taste of "big city" business. It helped develop his view of the importance of ownership of intellectual property. Westinghouse throughout his career would be a defender of inventor rights. The partnership continued downhill, and young Westinghouse began to look for alternative steel foundries to partner with. The search lead him to the Pittsburgh foundry of Cook & Anderson. They felt they could produce the steel frog at a lower cost and were willing to work with Westinghouse. George was adamant in wanting to leave his New York partners, and the patents were his. His concern was for Marguerite, who was extremely close to her family and friends in New York. Initially, the plan was for Westinghouse to go to Pittsburgh to assure a solid business, while Marguerite would remain awhile with his parents.

The Pittsburgh of 1868 was the industrial and metallurgical core of the nation. The Civil War had transformed the town into an industrial giant. Fort Pitt Foundry (formed in 1805) was one of the oldest iron foundries in the country, having supplied cannonballs to the Army of Andrew Jackson and the Navy of Commodore Perry in 1812. The Fort Pitt Foundry of Pittsburgh had been the major armory of the Grand Army of the Republic. It had pioneered casting techniques that allowed for the world's largest cannons to be produced. The famous monster Rodman cannon was produced there. By 1868, Pittsburgh was the major iron and steel producer in the nation. Pittsburgh produced about 45% of the nations total iron production. In 1867, Andrew Carnegie started his first iron works in Pittsburgh (known as Union

Iron Mills) and Benjamin Jones and James Laughlin established one of the first blast furnaces in Pittsburgh. Pittsburgh had six steel foundries and cast more steel than any other place on earth. By 1870 the sixty-eight Pittsburgh glassworks produced half of the nations total of glass. The growth in the glass, iron, and steel industries was related to its position in the nations greatest coalfields. The Pittsburgh firms of Graff, Bennett & Company and Sprang, Chalfant & Company were the first to apply natural gas to manufacturing in the 1860s. Pittsburgh was at the time the nations biggest oil refining area as well. It was also a natural transportation center, and was fuel rich having abundant coal and natural gas.

Railroads were a natural outgrowth of the booming industry, and Pittsburgh was a national center. The metropolitan area had over 200,000 people and was the fastest growing area in the nation. The first railroad chartered in Pittsburgh was the Pennsylvania Railroad in 1846. The track started in Pittsburgh and expanded to Turtle Creek in 1851 with a locomotive brought in by canal boat. A year later, the Pennsylvania Railroad was connected between Pittsburgh and Philadelphia. In 1855 the Baltimore & Ohio Railroad connected Pittsburgh to Baltimore as well as points west into Ohio. Pittsburgh being at the confluence of three great rivers (Allegheny, Monongahela, and Ohio) offered a natural merger point for river transportation with railroads. Pittsburgh represented an international port with a river connection to New Orleans and the Atlantic Ocean. By 1870 Pittsburgh's major employer was the railroad and related industries. The city had ten newspapers and over forty banks. Pittsburgh already had more foundries than any other city, but at the time it was known as the glass capital of the world. The rich coalfields of this area made Pittsburgh a natural industrial center. Most days found the city blanketed by a thick layer of smoke from the coal burning glass works, mills, and locomotives. It had already gained the reputation of a stormy sky-watch could turn the day to night requiring gaslights to see.

The city was known as a fortress of Republicanism having carried the elections for Abraham Lincoln. Its roots, however, were in Jeffersonian democracy that birthed the new industrial Republicanism of the time. Industrialists in Pittsburgh were said to have the status of today's movie stars and sports figures in the 1860s and 1870s. Pittsburgh attracted young industrialists like Hollywood does emerging stars today. The newspapers of the time idolized the city's industrialists and their triumphs. An aristocratic industrialist class was emerging in the suburbs of Homewood, Oakland, Shadyside, and Ridge Avenue in the city of Allegheny. Pittsburgh's North Side was then known as the City of Allegheny. Pittsburgh was a necessary stop for any presidential candidate as well as for the great entertainers of the day such as Mark Twain. The native population of Pittsburgh was Irish, Scottish, German, and English; it wasn't until 1880 that the first large influx of immigrants arrived. By 1867, a group of German iron makers were immigrating to Pittsburgh to manage the new mills. The "masters of Pittsburgh" were considered to be a unique blend known as the "Scotch-Irish race." Pittsburgh was also known as a bastion of Presbyterianism in 1860s, which was a formal extension of Puritan thought. Like New England of the time, Pittsburgh had a Puritan and Protestant work ethic as well.

George Westinghouse went by train to Pittsburgh in December of 1868 just as Mark Twain was finishing one of his famous oratories in Pittsburgh. He arrived at the city's landmark, the four story high Union Depot Building, which would later be burned to the ground in the Railroad Riots of 1877. December in Pittsburgh was a month of almost full "arctic" darkness due to the smoke; still Westinghouse seemed positively impressed by the industrial city. The city's streets were defined not by geometry but the geography of the hills and rivers making it impossible for visitors to find their way. It is a characteristic that modern visitors can still attest to today. It was this urban attribute that might well have helped determined Westinghouse's future success. Anderson and Cook was located on Second Avenue (Blvd. of the Allies) and Try Street. In his search for the firm of Anderson and Cook he got lost and asked directions, and met and formed a relationship with Pittsburgh industrialist Ralph Baggaley. Baggaley was a rising and prominent Pittsburgh businessman. Baggaley would be the future financier and manager of Westinghouse's air brake. After arriving at Anderson and Cook, he quickly closed a deal where Anderson and Cook would produce frogs and replacers, while Westinghouse would travel selling them. Marguerite remained in New York as George started a year of traveling and missing her. Living in cheap hotels, he saved all he could to move Marguerite to Pittsburgh.

Traveling to railroad executives allowed young George to work on a project he had been designing since 1867. That project was a new type of braking system for railroads. Westinghouse was a problem-solver, like Edison with the exception of a few pet projects; he focused on projects that would ultimately be highly profitable. Inventing represented a profitable career direction in Westinghouse's mind. His father's inventive efforts had inspired him. Westinghouse would have been well aware of the braking problems inherent to railroading in those days. Train wrecks were the subject of the newspapers and that of travelers. Westinghouse was on a train following a head on collision of two freight trains near Troy, New York in 1866. Westinghouse later would recount in a rare writing, his initial thoughts on seeing that collision:

> My first idea of braking apparatus to be applied to all of the cars of a train came to me in this way; a train upon which I was a passenger between Schenectady and Troy in 1866 was delayed a couple of hours due to collision between two freight trains. The loss of time and the inconvenience arising from it suggested that if the engineers of those trains had had some means of applying brakes to all of the wheels of their trains, the accident in question might have been avoided and the time of my fellow-passengers and myself might have been saved.
>
> The first idea which came into my mind, which I afterwards found had been in the minds of many others, was to connect the brake levers of each car to its draft-gear so that an application of the brakes to the locomotive,

> which would cause the cars to close up toward the engine, would thereby apply a braking force through the couplers and levers to the wheels of each car. Although the crudeness of this idea became apparent upon an attempt to devise an apparatus to carry the scheme into effect, nevertheless the idea of applying power to a train was firmly planted in my mind.[12]

Actually, Westinghouse tested the idea at his father's shop by building a model and making detailed engineering drawings. The small model proved unsuccessful. In any case, the need for and the commercial potential of a train braking system inspired many inventors such as Westinghouse. Westinghouse was on track with the idea of simultaneously applying power to all the wheels. Hand braking on trains was mechanically inefficient, manpower intensive, and extremely dangerous. The poor braking systems greatly restricted the length of the train to five to seven cars. The small number of cars on freight trains resulted in higher cost as well. Speeds over ten miles per hour were extremely dangerous. Braking required one to one and half miles of uneven stopping. Emergency stopping was non-existent, and even scheduled stops required a mile or more of planning. Track crossing in cities was a high-risk activity.

The hand brake system in place required coordination and muscle power. The locomotive engineer would start throttling down the engine as he signaled the brakemen with the whistle (known as the "down brakes whistle'). A brakeman was assigned between every two cars on the train. Ideally, the engineer had throttled to a coast as it approached the stop. The brakemen would begin the process of applying friction brakes on each car by turning a hand wheel, known as the horizontal wheel, on a vertical post at the end of each car that multiplied the force of his muscles. The strength required of the brakemen was the source of the term "Armstrong system." The hand wheel pushed brake shoes on each car against the wheels. The hand wheel used a chain to transfer the power to the brake shoes. After the brakes were applied to one car, the brakemen jumped to the other car to apply the brakes. Poor coordination between the brakemen could cause "locking" and serious jolting to the passengers. The system inherently caused a series of uneven jolts. Luggage was often thrown wildly inside the passenger cars. The danger to the brakemen was unbelievable. Jolting or locking often threw the brakeman off on his way across the car roof to the next car. The average life of a brakeman on the job was estimated at seven years. The brakeman had to race from car to car, often on roofs slippery with ice, snow, and rain. Even on the best days, the swaying, bobbing, and jolting of the cars caused a loss of footing. It was estimated a thousand brakemen were killed each year with another 5000 injured![13] While these deaths, as horrific and numerous as they were, didn't make the headlines, railroad wrecks were front-page news. In 1853,

12 Barbara Ravage, *George Westinghouse: A Genius for Invention* (Austin: Raintree Steck-Vaughn Publishers, 1997)
13 *The Search for Safety*, Union Switch and Signal Division of American Standard, Pittsburgh, 1981

President Franklin Pierce had a son killed, and even Charles Dickens had been in a major brake failure wreck in 1865. Dickens, like many of the time, was fearful of rail travel. A 1856 head on crash in central Pennsylvania had killed over fifty people, mostly children on their way to a Sunday picnic. The government was also taking note of the problem as the public fear and outcry increased.

Westinghouse's job of selling frogs and replacers for Anderson and Cook allowed him to discuss and work out problems with a future brake design. On a sales trip to the Chicago, Burlington, and Quincy Railroad, he saw a trial of a different braking system. Augustine Ambler of Milwaukee had patented this system in 1862. The system had been adapted to the Aurora Accommodation of the Chicago, Burlington, and Quincy Railroad. The Ambler was a "continuous chain brake." A chain connected the friction brakes on each of the cars to the locomotive; the Aurora Accommodation was pulling five cars at the time. By winding up the chain at the engine using a windlass, the brake shoes were closed on the individual cars. Ambler used the power of the locomotive's drive wheel to turn the windlass. The chain system worked when limited to five cars and was very costly. Passenger trains of the time were already approaching lengths of ten to twelve cars. Amber had only convinced one railroad to experiment with his device. Westinghouse started to look at ways to improve on the chain system. The key he believed to adapting it to longer trains would be a more efficient application of the locomotive's power. His first idea was to use the engine's steam to drive a separate piston to pull the chain. He worked with models of this new system, but concluded through his calculations that the steam cylinder would be larger than the locomotive itself to pull for a ten-car train. Most of this modeling and drafting was done at his father's shop. All this as he worked to sell frogs and save money to move to Pittsburgh.

To deal with longer trains, his models suggested he needed to have a cylinder on each car to operate the brake shoes. Some of the work had resulted from his original model. Modeling was an engineering tool that Westinghouse perfected over the years, passing the technique on to generations of future engineers. The idea would be to use steam power. Steam would be supplied to the individual cars by a long pipe with the necessary flexible connections between cars. A steam driven cylinder could then activate the brakes on each car. Again Westinghouse's modeling and experiments suggested that steam would condense in such a long transmission pipe, particularly in cold weather. The modeling did show that if he could avoid condensation, steam could be used to transfer power to the brake shoes. Westinghouse was constantly talking to railroad engineers and executives in his travels. He was well aware of a number of inventors in England and America working on the same problem. Disappointments continued to mount with the steam to activate individual brakes; a new type of power transmission would be needed.

The idea came from a magazine detailing the building of the Mont Cenis Tunnel in Europe. The Mont Cenis Tunnel was to be the world's longest, at eight miles. It would connect the railroads of Italy and France, following one of Napoleon's routes through the Alps. The project was planned to take 20 years because of the short working seasons in the Alps. The project started in 1858, using hand drilling which resulted in a rate of only about a third of a mile a year.

J. Couch had invented steam drills in America in 1849. They were being used in the drilling of very short tunnels in New England. In 1861, Thomas Bartlett had invented a steam drill for drilling coalmines. The problem with using steam drilling at Mount Cenis was the same as Westinghouse's brake transmission problem; the steam condensed, limiting the distance a pipe could carry it. The steam engine could of course be moved into the tunnel as the drilling progressed, but the coal burning to produce steam fouled the air. Bartlett started to experiment successfully with compressed air on the Italian end of the tunnel, compressed air being a form of stored energy that could be delivered to the point of need. The steam engine was placed outside as compressed air was delivered over 3000 feet. Compressed air drilling progressed at four times the speed using an 8-inch along the wall to transmit power. By 1868, the compressed air tooling was making news as new tunnels were being planned all over the world. Westinghouse had been a vigorous reader of the technical and engineering journals from his earliest days at his father's plant. The Mont Cenis Tunnel had become a weekly report in the journals. Westinghouse immediately saw the application of compressed air to railroad braking, but he was not alone.

In early 1868, Westinghouse had engineering drawings and models showing the practicality of using compressed air. The suggestion of compressed air in railroad braking was not unique to Westinghouse, but a complete engineered system was. By June, Westinghouse applied for a patent of the idea before he had an actual test based on his models. This would eventually give him the edge in several patent battles that occurred early on. This first Westinghouse brake was known as the straight-air brake. It was simple and primitive. A steam pump on the locomotive using the same steam that drove the engine generated the compressed air. The compressed air was contained in a reservoir located on the locomotive as well. A pipe using flexible couplings was used to connect the individual brake cylinders on each car. These flexible couplings were three-ply rubber, which was available and could handle compressed air (but not the steam of Westinghouse's original idea). When the engineer released the compressed air in the reservoir, it rushed to the individual cylinders, which closed the brake shoes on the car wheels. It used friction like our modern car to stop the car. Of course in today's car, hydraulic fluid is used to apply the force on the brake shoes. The action was solely controlled by the engineer eliminating all the brakemen and applying the brakes evenly and simultaneously. The summer of 1868 was one of continual activity for Westinghouse. On his sales trips to sell frogs for Anderson and Cook, he tried unsuccessfully to promote a trial of his air brakes. The board of directors of the Chicago, Burlington and Quincy were the first to turn down the trial. It was a disappointment because he had successfully convinced the superintendent that the railroad of the application of the air brake would work. The next and more famous rejection was from Commodore Vanderbilt of the New York Central line, but he did convince the New York Central to buy his frogs, generating some badly needed cash.

The Vanderbilt story has achieved legendary status among the Westinghouse stories. On the day of his death the story was reported in over hundred obituaries across the nation and world. The story goes that in 1868, he visited the office of Commodore Vanderbilt, then president of the New York Central Railroad. It was officially a visit about selling him railroad frogs, but Westinghouse intended to pitch his air brake as well. Vanderbilt, not knowing Westinghouse at the time allowed him to wait as other salesmen. Westinghouse tried for two days, and was finally granted a meeting. He presented his frogs, Vanderbilt had heard of, then concluded with a sales speech on his proposed air brake. Vanderbilt bought the frogs, but responded to the air brake proposal with, "If I understand you, young fellow, you propose to stop a railroad train with wind. I'm too busy to listen to such nonsense."

After the major sale of frogs to the New York Central, he finally had saved enough to bring Marguerite to Pittsburgh in July of 1868. He would rarely travel without her from then on. Marguerite's coming to Pittsburgh energized George again. For Marguerite, it was a real challenge. Pittsburgh was a man's town at the time. The dust, smoke, and dirt made wearing white or light colored clothing impossible. Home decorations were also dark to hide the constant dirt, which added to the dark and gloomy environment. Before 1880, Pittsburgh had no parks for recreation. George and Marguerite started out in a rented flat on Pittsburgh's North side (then known as Allegheny City). It was a difficult start, since the North Side of Pittsburgh was a tough area. Irish and Scotch gangs often clashed. In the meantime both he and his wife developed a friendship with Ralph Baggaley, which would bring them into Pittsburgh's better society. Baggaley was from a prominent family and was managing one of the city's best foundries. Baggaley's interest in the air brake had grown out of the friendship. He offered to invest a few thousand dollars to help George build a working system that could be tested on a train. George hesitated to involve his new friend in this risky investment that even his father had balked at. He allowed Baggaley to have a consultant look at the investment. The report came back negative, calling it impractical. Baggaley, however, wanted in, and formed a handshake partnership to build experimental apparatus. Baggaley's foundry did not have the right machining equipment, so they contacted the firm of Pittsburgh's Atwood and McCaffrey to make parts. The firm was supplied with Westinghouse's detailed drawings, which made production simple and standardized. George used all his spare moments between sales trips to supervise the production of brakes at the factory and work out operational and design problems. Baggaley, in the mean time, worked on setting up interviews with Pittsburgh railroad executives.

One of those meetings was with Robert Pitcairn, local Superintendent of the Pennsylvania Railroad. Pitcairn was and had been since boyhood friend of Andrew Carnegie. He had even invested with Carnegie in a number of coal mining ventures. Pitcairn loved the idea and Westinghouse, but he could not approve a trial. Pitcairn brought in the superintendent and assistant superintendent of Altoona, Edward Williams and Andrew J. Cassatt. The meeting was a setback with the Pennsylvania Railroad refusing to fully finance the trial. One problem was Westinghouse's youthful appearance, at the age of twenty-two. Still, they couldn't help but be impressed. The next meeting with W. W. Card, Steubenville Superintendent of the Panhandle Railroad, was successful. Card went with Westinghouse to Atwood and McCaffrey to go over each part of the design. Card was almost wildly enthusiastic about the brake, but his board also turned down the financing. Card had an excellent view of manufacturing from his purchasing responsibilities at the railroad. Now, it was Card who helped Westinghouse and Baggaley to arrange for a loan. Card would ultimately be one of the first investors in Westinghouse Air Brake. Finally, it was decided that a train would be equipped with the Westinghouse system in September of 1868.[14] This date has been confused with April 13, 1896, the actual date that the patent was issued.

The Steubenville Accommodation of the Panhandle Railroad was equipped for the test at Baggaley's North Side plant, mostly by Westinghouse himself. The trial was with four cars. The last car on the train would carry some special quests including Card and some of the Panhandle directors. Daniel Tate was the engineer and Westinghouse had trained him personally on how to use the brake. Westinghouse gave Tate fifty dollars as the train left Panhandle Station in Pittsburgh on its fifty-mile trip to Steubenville. Westinghouse went to sit in the locomotive cab. Normally accommodation trains went about a quarter mile through the Grant Hill Tunnel and stopped at Fourth Street to pick up passengers. This reserved train, however, was not scheduled to stop in Pittsburgh that day. It crossed Second Avenue and the Monongahela Bridge and then open track following the Ohio River. The train left the tunnel at thirty miles per hour, only to see a merchant and his cart on Second Street. It is not clear where the planned stops were that day, but this was an unexpected emergency requiring Tate to use the brake for the first time. He pulled the hand lever quickly, bringing the train to a jarring stop. The guests in the last car were thrown from their seats in confusion. This first successful trial is immortalized in a painting today and is verified in personal letters of the engineer and superintendent Card. Having had such a spectacular demonstration, there was no question about proceeding, which they did, with other tests along the way.

14 Prout, 29

Westinghouse telegraphed his father of the successful test, but there was no publicity or even a mention of the test in the many Pittsburgh newspapers. At thirty miles per hour, hand braking if done well could stop a train within 1600 feet. The Westinghouse "straight air brake" could do it within 500 feet. Westinghouse knew he needed to a lot more work on the system. It still had some major drawbacks; one was that it was not a fail-safe system. The compressed air reservoir was in the locomotive. When the engineer released the compressed air it rushed down the train, activating and powering the brakes. The time to get the compressed air down the train at the needed pressure increased with the length of the train. The pressure also decreased down the length of the train reducing the force applied. If a coupling broke at any point, air and pressure were lost and no braking occurred. This was a major concern because without a fail-safe design, the railroads could not eliminate the manual brakes and brakemen. Furthermore, the Panhandle Railroad system was a piece of craftsmanship; in the future, Westinghouse would need to standardize his parts and assembly to make it a commercial success. Westinghouse would eventually resolve all these problems, but his immediate concern was to get new trials and manufacture more systems.

Word of the trial did travel in local railroad circles quickly. The railroads, however, were cautious. First, brakes to many railroad executives were a safety issue, not a profit issue. In fact, Westinghouse's success struck fear in many because of the cost of potentially converting all existing trains. One estimate was that $40 million would have been needed in 1875 to convert the industry (more like a billion in today's money). Pitcairn and the Pennsylvania Railroad were again interested in a new trial, seeing an opportunity for personal investment if nothing else. Westinghouse and Baggaley now had railroads and even their executives willing to at least invest in trials. Baggaley started to look at the possibility of producing the apparatus at his foundry. Also Westinghouse still had not received the formal patent, which slowed things a bit as well. Still, Westinghouse resigned from Anderson and Cook to devote full time to the development of commercial equipment. When the patent was approved on April 13, 1869, both Westinghouse and Baggaley were ready for production. Trials on a six-car Pennsylvania Railroad train were readied.

Early 1869 was indeed a busy time as Westinghouse saw an increase in interest. The six-car Pennsylvania Railroad was being used for exhibition as it transversed the Allegheny Mountains between Pittsburgh and Altoona. Many of these trips included reporters and articles started to appear across the nation. In September of 1869, this special train was on display at the Annual Meeting of Master Mechanics from all over the world got to ride on it. In November, a ten-car Pennsylvania Railroad train was fitted with the Westinghouse apparatus. This train went on to Philadelphia. These demonstrations were repeated in Chicago, Indianapolis, and St. Louis. These tours were the brainchild of Westinghouse, who was selling the railroads through the public. Reporters were always on the VP quest lists in each city.

Westinghouse was facing a tough sales battle. Railroads while impressed, now balked at the cost of converting to the air brake system. The New York Central of Vanderbilt remained in major opposition. The hand brake manufacturers banded together with a negative advertising campaign to counter Westinghouse's success. Westinghouse's sales experience paid off, however, with his publicity tours. By 1870, Westinghouse had firm orders from the Michigan Central, Union Pacific, and Chicago & North Western, as well as many of the Eastern lines.

Chapter 4. Pittsburgh Industrialist

In July 1869, Westinghouse Air Brake was formed with a board of directors from the enthusiastic railroad executives (the same year H. J. Heinz opened his first plant in Pittsburgh). The board consisted of a twenty-two year old Westinghouse, Robert Pitcairn, W. Card, Andrew Cassatt, Edward Williams, G. Whitcomb, and Ralph Baggaley. The capitalization was $500,000. Cassatt, Pitcairn, and Williams were Pennsylvania Railroad executives, and Card and Whitcomb were from the Panhandle Railroad. John Caldwell, a friend of Baggaley was elected treasurer of the new company. By the end of 1869, the nation was aware of the new Westinghouse brake, and its potential to improve railroad safety. The Westinghouse train tours had been a huge success with the public, still the cost was holding back some of the railroad executives. The major lines already had a massive amount of capital invested in braking systems, and the Westinghouse system was not fail-safe, even though it offered a significant advantage over hand braking. Public pressure was building, as accidents seemed to fill the headlines. Westinghouse started to focus on the nation's mechanics and railroad engineers through the scientific journals to increase interest and sales. It was clear that the railroads could only delay, not prevent the conversion of the nation's railroads to the air brake. Still there were major challenges; in particular, the straight air brake was limited to ten cars because of the decrease in air pressure by the end of the train. In 1870, trains were moving to twelve cars, so Westinghouse had to look to the future. The smaller railroad lines were ordering the system because they used it for a competitive advantage of safety and speed. The small lines also tended to run "accommodation," which meant short runs and stops every few minutes through populated areas. Going into 1870, Westinghouse Air Brake was struggling to meet the orders of a booming market.

In early 1870, Westinghouse Air Brake leased a building at Liberty Avenue and 25th Street in Pittsburgh (Strip District). This area was at the heart of the Pennsylvania Railroad operations. Westinghouse and Baggaley were now fully immersed in the production of air brakes. Baggaley functioned as the operations manager having had experience in managing a foundry, freeing Westinghouse to focus on the engineering and sales. Westinghouse applied all that he had learned in his father's Yankee machine shop. One policy he set went back to the philosophy of all of the great New England mechanics of the early 1800s-stanadrdization. Eli Whitney had shown its power to Thomas Jefferson in 1812, and New England's Springfield Armory had spread its benefits to the production of armaments. Another New Englander, Thomas Blanchard, had applied it to machine and tool making. Standardization was far form a general industrial practice in 1870. Standardization was critical to railroad maintenance operations. Westinghouse became the apostle of standardization to the point of risking easy entry of competitors. In fact, often inventors steered away from standardization in order to avoid copying by competitors. The strategy of non-standardization is often used today to avoid copying, but often have disastrous market results. Apple Computer used it for years until the interchangeable, cloned IBM overwhelmed the market. Sony tried it with the technical superiority of beta recorders, only to lose the market to the inferior but standardized VHS recorders. Westinghouse had learned well in New England the power of standardization in selling and repairing agricultural equipment. He realized the huge railroad industry would need standardization to facilitate repairs throughout the world. Of course, Westinghouse believed in intellectual property and the protection afforded by the U.S. Patent Office to prevent copying. Westinghouse took standardization further than any other private industry had with the mass production of air brakes. To that end, Westinghouse applied for patents on the key components, such as the steam engine apparatus to compress the air. Standardization was a philosophy that Westinghouse applied throughout his career.

Westinghouse was overworked in 1870 in trying to get production up, sell the big railroad lines and continue developmental work toward a fail safe, automatic brake system. His research into competing patents of his brake led him to conclude that the British patents of air brake systems, showed that these European systems lacked performance. The British system did not use a reservoir of compressed air and therefore were slow to react as the compressed air was generated. It was clear that the Westinghouse straight air brake offered a competitive advantage over any thing available in Europe. Westinghouse's pump/reservoir was unique and he applied for patents for all parts of it. More importantly, Britain offered huge sales potential for his air brake. With Baggaley as a strong production manager, Westinghouse was free to focus on the conquest of Europe. In July of 1871, Westinghouse and his wife made their first trip to Europe. Because of the nature of travel, Westinghouse planned to stay until August of the following year. Marguerite was determined never again to be separated from her husband. It would be one of many trips to Europe they would take together. They were almost inseparable as a couple, and when George was in Pittsburgh, he always left work at 5:00 to have dinner with Marguerite (often bringing home a number of associates). Marguerite was his best friend throughout life. In addition, they traveled often back to New York to see family. Later in the decade Marguerite was one of the first in Pittsburgh to have a telephone, and George got into the habit of calling her every day. Besides, money was flowing into Westinghouse Air Brake as the smaller American railroads were rushing to get the system. He had even purchased a large house in Pittsburgh's fashionable east suburb of Homewood. He surprised Marguerite on her birthday in 1871 with their new home. Marguerite had struggled with the smoke filled air of industrial Pittsburgh. The Homewood area was heavily wooded and offered a break from the dirt and smoke of downtown Pittsburgh. He would call this beautiful estate, "Solitude," reflecting the importance he put on having a home in which to relax. Westinghouse always stressed "home," since being with Marguerite was of the utmost importance.

Westinghouse prepared a detailed strategy for selling Europe with the help of his Board of Directors. Those critics of Westinghouse's business in the Panic of 1907 overlooked his real business genius. Westinghouse was a master of strategy. He prepared a detailed sales pamphlet to be mailed to Europe's railroad executives prior to calling on them. This again was a tactic, he had learned from his Yankee machine shop roots. He also focused on articles and news releases for the scientific and engineering journals, because his straight air brake had a technical advantage over European systems. Still, he faced a cold reception and a tough sell. The British-American relationships of 1871 were like France and the United States today. The British maintained an air of superiority and tradition. American railroads were seen as risky moneymakers owned by the "Robber Barons" and cowboys. The British wanted a fail-safe system before they would convert. This was a much different approach to engineering than that in America. Engineering in America was more evolutionary. His visit to the Mr. Dredge, editor of *Engineering*, typified the trip. Dredge had been leading the charge for better railroad brakes in England. While Westinghouse and Dredge hit it off personally, they had disagreements on the brake apparatus. Dredge was a strong believer in a fail-safe system. He pointed out that a failure of the reservoir and/or car pipes or couplings would result in no braking. Westinghouse faced this stiff criticism, even though; no present system in England was truly fail-safe. While disappointments, Westinghouse had learned as he had in the past through his discussions with railroad executives.

Westinghouse was given a set of requirements spelled out in detail from the engineering journals and societies, and these would become the design criteria for future Westinghouse designs. The British approach to design varied from the American. The British engineers looked at how a braking system might fail and then set necessary requirements. Eight points were made:

1. Brakes should have the ability to be applied not only by the engineer, but the guards on various cars. This was much different from the engineer operated system of America.
2. The application of the brakes should require only a small exertion versus the strong-arm requirement of brakemen.
3. Application of the brakes should allow for a slow braking as in approaching a station or for an emergency stop, such as a blocked crossing.
4. If cars broke away from the train, brakes would automatically kick in. This would require some future work by Westinghouse.
5. The system should allow easy coupling and decoupling of cars.
6. Brakes on each car could be applied independently. Westinghouse would resolve this in 1879 with reservoirs being put on each car.
7. Brake failure on any one car would not interfere with brake operation on other cars.
8. The system should be reliable and easy to maintain.

These requirements become the Magna Carta of European brake design. Years later Westinghouse was thankful for the technical challenges laid down by Dredge. Westinghouse would achieve all these requirements over the next ten years, giving him a worldwide advantage.

Dredge was right in that Westinghouse's system had a reliability problem; even the converted American trains still had brakemen in place for emergencies. Westinghouse's solution to the reliability problem demonstrated his engineering background. He reviewed his system as to the places that had a high probability of failure and addressed them. The solution was engineering redundancy that we see applied in today's space program and nuclear plants. He designed in redundant car pipes (a double line of pipes) and couplings, and he also added a back up compressed air reservoir to the engine. The system could function as one or independently via the use of valves. He applied for patents in December of 1871 for both these designs (they were granted in 1872). Westinghouse's own words describe them best:

> In the air-brake apparatus heretofore in the use a single line of pipe conveys the compressed air from the main reservoir on the locomotive to each brake cylinder. If this pipe becomes accidentally broken at any point it is, of course, useless for braking purposes from that point to the rear end of the train. For this and other reasons I have devised an apparatus consisting in part of double line of brake pipes, which may be cooperative or independently operative in braking, at the pleasure of the engineer. . . The improvement herein described consists in features of construction and combination by which, first, an air reservoir, auxiliary to or independent of the main reservoir, is combined on each car with a brake cylinder.

The 1871-1872 European trip served well to start Westinghouse as an international manufacturer. First he hired a British business agent to help develop his European business and stay in touch with the railroads. Westinghouse spent a great deal of time studying European technology. It was surprising to find the British using steel rails (years ahead of America), yet using wood brake shoes instead of steel or iron. Furthermore, Westinghouse took the time to apply for British patents and learn their legal system. Before he left England, he was able to get several key trials. One was with the North-Western Railway Company on its line between Strafford and Crewe, which proved successful. He was able to get a trial on a 12-car train of the Caledonian Railway line between Glasgow and Wemyss Bay. This trial was also successful, but both lines were heavily invested in chain brakes and hand brakes, and they were reluctant to change to a new system. Marguerite and George also took the time to visit France, Germany, and Belgium. Westinghouse started to manufacture these suggested British improvements in later 1872, but prior to this, another event in the United States took place that caused sales to boom.

That event was the crash of New York Central's *Pacific Express* on a wintry night in February 1871. The Pacific Express was Commodore Vanderbilt's favorite and had been dubbed "America's Number One Train." This day, it was making its run from New York City to Chicago. Approaching Wappinger Creek Bridge near Poughkeepsie, New York, the engineer saw an obstacle on the bridge and signaled "down brakes." A southbound freight train of twenty-five oil tank cars had jumped the track due to a broken axle. The tank cars lay all over the bridge. The "down brakes" whistle was too late to stop the *Pacific Express*, and it ploughed into the tank cars, resulting in fiery explosions and the ultimate collapse of the bridge. Thirty were killed with many more injured, making headlines across the nation. The accident jolted Commodore Vanderbilt into action, who had earlier thought little of Westinghouse's air brakes. In 1872, the New York Central ordered brakes for all its passenger trains. The conversion of Vanderbilt's railroad caused a stampede in orders for air brakes. The market was now wide open for Westinghouse who was fast becoming a tycoon. Still Westinghouse worried about a similar disaster if an air brake were to fail; he was not yet meeting the British technical challenge to assure reliability. His upgrading of the system using redundant piping and individual car compressed air reservoirs was one step towards that goal. Still, the Westinghouse Air Brake was limited in all emergency full stops. Westinghouse was every bit the true engineer worrying and wanting perfection. He had learned to build his designs based on potential failures. This type of problem solving design would become the hallmark of Westinghouse and one of his major contributions to the field of engineering. Again, Westinghouse's absolute demand for reliability in his products and systems can be traced to the machine shop methodology of New England. This engineering approach is today known as Failure Mode Effect Analysis, and is a popular engineering discipline in transportation and defense industries.

Trains were subject to two very distinct modes of stopping. A service or station stop, at which the engineer slows down and the brakes are applied at less than full power. The straight air brake applied air pressure to close the brake shoes onto the car wheels. The slower type of service stop allowed plenty of time for full air pressure to develop throughout the train. The second type was the emergency stop requiring quick action and full power. This quick braking taxed the ability of the system to apply full pressure the length of the train. Also, ruptures could disable the system. Leakage and mechanical failure were the main failure modes of the straight air brakes during emergency stops. The use of duplicate redundant systems helped reduce the possibility, but did not improve brake responsiveness. In addition, the length of trains was increasing every year straining the power that could be applied directly by compressed air. Westinghouse began to feverishly work on a system that would automatically apply the brakes even in the face of a mechanical failure, which was another of the British eight points of design. His first approaches included beefing up the system with better parts and steel, but he needed a radical new design to approach the goal of fail-safe. Westinghouse had extensive models at his plant to study new systems. By 1872, he had a working model of a new system. In 1873, Westinghouse started to add a supplemental and independent system to trains that would eventually evolve into the "automatic brake."

Westinghouse's idea of the automatic brake represented his true engineering brilliance. The automatic system reversed the use of compressed air. In the straight air brake compressed air supplied the force to put the brake shoes against the wheel. The loss of pressure meant no force could be applied. The automatic system used compressed air to keep the brake shoes off the wheel. If pressure were lost due to leakage, a rupture, and/or uncoupling, the brakes would automatically be applied. The automatic action was achieved by what has become known as the "triple valve." The triple valve eliminated the need to constantly supply air pressure to keep the brake shoes off the wheels. The triple valve compressed air was used only to send a signal to the brake shoes. The triple valve acted as a controller or manager of the system. It was a triple valve because it managed three modes of operation-charging the car's reservoir with compressed air, applying mechanical pressure of the shoes with a reduction of air pressure, and releasing the brakes with an increase in pressure. This resolved the problem of maintaining high pressure along the length of the train. With his pumps and individual reservoirs, he had enough air pressure for miles of cars. This triple action allowed for gradual application and releasing of the brakes in proportion to a change of pressure by the engineer.

With the inside help of his directors, Westinghouse was able to get a trial with the Pennsylvania Railroad in May of 1873. A joint committee of the Pennsylvania Railroad and the Franklin Institute monitored this exhibition train. Tests were run on this seven-car train in the Philadelphia area. The train demonstrated the ability to stop in 15 seconds within 553 feet at a down hill speed of thirty miles per hour. These results startled the engineering world and resulted in the first of many medals and awards. The committee report concluded:

> The committee say that these experiments have demonstrated to them the extraordinary efficiency of this apparatus, and they especially call attention to the value and importance of the arrangement which secures the instant automatic application of the brakes on the engine and on each car of the train independently of the train hand, in certain contingencies which are of common occurrence and are the cause of frequently disastrous accidents. The committee believes that by contriving and introducing this apparatus, Mr. Westinghouse has become a great public benefactor and deserves the gratitude of the traveling public at least. They believe that his inventions are worthy of and should receive the award of the Scott Legacy Medal.

The success and national publicity spurred sales to even newer heights. Westinghouse was engrossed in making major improvements to his system to meet the European competition. This newly designed system would be known as the automatic brake but with greatly improved shoes, beams, and control. He continued to design backwards from potential modes of failure. The approach improved reliability, brake life, train weight, and train speeds. He continued to pioneer the use of steel. He needed its high reliability to eliminate the need for back up hand brakes and brakemen. The reduction of brakemen was one area that would help reduce the cost of the Railroads in converting to his system. Another issue was that while the railroads were adapting the straight air brake to passenger trains, they hesitated using the system on long freight. Engineering demonstrations, exhibits, and trials were almost a passion for Westinghouse the engineer. Money meant little to him in comparison; engineering achievement was always his goal, not capital gain. It was a passion that would come back to haunt him in the Panic of 1907. The Pennsylvania Railroad demonstration showed that Westinghouse could outfit a long train, and his brakes could stop any train in the world.

One of the side benefits of the improved brake was a new interest in steel. Westinghouse pioneered the use of steel in his steel frogs and replacers. Steel had increased the life of these products ten-fold. Westinghouse's fascination with steel and his repeated success with it had caught the eyes of many railroad executives. England had been converting to steel tracks in the late 1860s, but in 1873 American tracks were either wrought iron or cast iron. Since frogs were basically a piece of track, Westinghouse's frogs had clearly made the case for steel, but railroads were slow to convert because of a lack of Bessemer steel mills. Other than Alexander Holley, there was no bigger promoter of steel than Westinghouse. Westinghouse had seen the Bessemer steelmaking process in his 1871 trip to England. Westinghouse even started to experiment with different types and compositions of steel far away from Pittsburgh steel master Andrew Carnegie. As late as 1872, Andrew Carnegie was not convinced that steel had any role to play. In fact, it would take his brother Tom to push him into the business. One Westinghouse convert to steel was the President of the Pennsylvania Railroad, Edgar Thomson. Edgar Thomson continued to test steel tracks with similar results in wear that Westinghouse's frog had shown. In 1875, Andrew Carnegie opened the world's largest Bessemer steel mill to produce railroad track in Braddock, Pennsylvania, naming it after Edgar Thomson.

Just as revolutionary, as the emergence of Westinghouse's use of steel and his discipline of engineering was his employee practices. After his first trip to Europe, he returned and introduced a five and a half day week. He established this in a time when they were having difficulties filling orders. Having a half Saturday off for family and Sunday for God was unheard of in industrial enterprises of the time. The practice was traceable to the shop of his father in New England. The year before he had instituted his employee Thanksgiving Dinner at the Union Hotel. The Thanksgiving dinner include the whole family and showed the importance he put on family. George and Marguerite sponsored family picnics in the summer as well. Westinghouse did these things from his heart, but many later claimed the motive was to increase productivity. To Westinghouse, the increase in productivity was a side benefit of doing the right thing.

In any case, productivity did boom through good management, good engineering, and a caring approach to his employees. In 1871 the plant at Liberty and Twenty-Five produced 18 sets of car brakes equipment and 4 sets of locomotive brakes per day using 105 people. By 1880 he could meet the demand producing over 100 sets of brakes a day. By then he had instituted another revolutionary employee benefit, the paid vacation. Employment had grown to over a thousand, but he was manufacturing a wide array of equipment as well. By 1890 at his new plant in Wilmerding, he was producing over 1000 sets of brakes per day as well as assorted railroad equipment, using 6000 employees.[15] Westinghouse never changed his deep concern for the employees. Westinghouse never saw himself as anything other than a gifted representative of God, while other great industrialists, such as Carnegie viewed themselves as trustees of God. People came first because that was how he was brought up in New England. He sincerely cared about his fellow man. His fellow men's welfare was one thing that went ahead of his huge desire to achieve. The reader today may find this view of Westinghouse too heroic or revisionist for reality, but Westinghouse was every bit a kind and generous man, even to heroic levels. When Westinghouse Companies asked for stories of Westinghouse by employees for the1930 anniversary, letters poured in of stories of his kindness and help.

The impact of the air brake on passenger trains was noticeable by the new role of the "brakeman." That role was highlighted in the 1874 August issue of *Harpers Magazine*:

> Since the invention of the Westinghouse air brake the office of the brakeman has sensibly decreased in importance . . . It places the whole train under the engineer's control. The through fast express trains on our great

15 Frank Crane, *George Westinghouse: His Life and Achievements* (New York: W.H. Wise & Co., 1925)

> trunk roads are now, we believe, generally supplied with this contrivance. But the train cannot spare the brakemen. He stuffs fuel into the stove at the request of the passenger that is too cold, and opens the window at the request of the passenger that is too hot. He unlocks the seat and turns it over for the mother who wants to convert it into a lounge for her tired child to sleep on. He opens the door and shouts in stentorian tones unintelligible words at the approach of every station. He occasionally makes announcements, but as he usually does this when the train is in full motion, and he has never been taught to articulate very distinctly, the passenger who is curious to know the meaning of his address has always to ask for its private repetition. He is always on hand to help passengers off the platform. Of men he is decidedly oblivious; he is a ladies' man, and the assiduity of his attentions is generally in direct proportion of their youth and beauty. When he can inveigle a young lady on to the platform before the train has quite reached a stop, and can protect her from falling by gently encircling her waist with his strong arm, he is perfectly happy. A virtuous brakeman is never without his reward.

In 1874, Westinghouse brakes were on 20 percent of the nation's trains. All of these were the straight air brake. Production of the automatic brake started in 1874, and replaced the straight air design. The automatic brake system addressed several of the British design concerns. The straight air had a major problem in that a break in car couplings, an air line leak, or a train break would render the entire system useless. To remedy these deficiencies, the automatic brake system had compressed air reservoirs on each car to allow each car to function independently. In addition, the "triple valve" allowed a fail-safe condition that applied the brakes when anything disengaged the overall connections. Clearly, Westinghouse was driven to meet the eight demands that had blocked him in England. In 1876, he had 37 percent of the existing trains or 2,645 locomotives and 8500 passenger cars. By 1879, Westinghouse had installed his system on 3,600 locomotives and 13,000 cars, and mainly these were automatic brakes. With the booming sales of the automatic brake, Westinghouse was in need of good mechanics and machinists. Straight hourly wages for such skilled employees were around two dollars and fifty cents per hour. Westinghouse devised a piece rate system based on units or work done. Employees actually earned four dollars an hour. Westinghouse preceded the famous Fredrick Taylor piece-incentive system by twenty years! Westinghouse again brought his unique vision of the industrial workers as a new production type craftsman. Machinists in his father's shop had that type of status. Industrial Pittsburgh had evolved into the industrial revolution with the iron and glass industries, which looked at men as a pair of hands to do a specific job function. Pay was by the hour and job type. It dehumanized the work, where Westinghouse heralded a different type of industrial worker. These new industrial craftsmen were paid based on knowledge and performance. This vision was a century ahead of its time. Westinghouse loved to be in his shop talking and taking the feedback of his employees. Often he would incorporate these ideas and pay them for them. They worked hard, got paid well, and had weekends off things few workers had in industrial Pittsburgh. Employment at Westinghouse was highly prized and the envy of area mill workers. The safety performances required by Westinghouse devices demanded a worker's best effort.

I. Levine noted, "In a single decade between 1870 and 1880 the accident rate was slashed so dramatically that the American public lost its fear of traveling by train." He still was stalled in Europe as seen. His first trip had won him some trials like the Caledonian Railway between Glasgow and Wemyss Bay, but the British railroad system was highly regulated by the government. Sales depended on endless trials and government reports. His automatic brake system had taken steps towards satisfying the British engineers, but he would need to promote daily. In 1872, Westinghouse had made a goal and commitment to supply Europe; he formed a brake export company with offices in England. Marguerite and George returned to England in 1874 to introduce the automatic brake, which corrected the 1871 criticisms of his straight air brake. England was in the middle of what would become known as the "war of the brakes." In the mid-1870s, there were numerous competing systems including air brakes, chain systems, and vacuum systems, with the major competition coming from the vacuum brake.

Westinghouse had actually invented a type of vacuum brake in 1871, but Westinghouse only improved on an 1860 patent of Nehemiah Hodge. Also British meta-inventor James Nasmyth had patented a vacuum brake in 1859. The Westinghouse version combined the vacuum on one side of his air brake to quicken application of the brakes. Interestingly, the vacuum brake competition in England was from another Pittsburgher, John Smith, who patented a direct vacuum brake. Smith's brake had found no success in America because of the success of the air brake. England's government and regulating committees and the early popularity of vacuum brakes had slowed all progress of air brakes in Europe. Typical of the approach, a grand series of trials were proposed for 1875. In May of 1875, Westinghouse again returned to England for several weeks to observe these trials. The trials became known as the "Newark Trials" because they took place at the Newark Division of the Midland Railway. These trials would include Fay's hand brake, Clark & Webb's chain brake, Barker's hydraulic brakes, and Smith's vacuum brake. It was the first battle of the brakes and Westinghouse would win. At fifty miles per hour the stopping distances were 777 feet for Westinghouse's automatic brake followed by 901 feet for the hydraulic system, and 1477 for the vacuum system. The success of these trials was far from the end of the war. Vacuum brakes continued to lead in sales. Westinghouse did win over the editors of *Engineering*, who reported in June 25, 1875:

> Lastly, we come to the Westinghouse automatic arrangement, and this, we think we may safely say, is shown by the recent trials to possess all the requisites of a thoroughly efficient continuous brake. This brake proved more prompt and powerful in its action than any of its competitors . . . Its performances as they stand were far beyond those of any other brake. As regards durability and general reliability in every-day practice also it should be remembered that no brake sent to the trials has been so thoroughly tested as the Westinghouse, and this is a fact which it is well to bear.

This editorial was a source of great satisfaction for Westinghouse, who in 1871 had been put back by harsh criticism of his straight air system. Yet, the British Railways continued to favor the Smith vacuum brake. Westinghouse remained optimistic, and set up a production facility outside of London. Some of the resistance of the British Railways resulted from their earlier commitment to the Smith vacuum brake. Conversion after such a large investment to another capital-intensive system was a significant barrier. The same barrier of course had worked to his advantage in the United States were he dominated early on.

On Westinghouse's return to America, he started preparation for the Centennial Exposition of 1876 in Philadelphia. This Philadelphia World Fair would be used to highlight the Westinghouse Air Brake to technical people from all over the world. Westinghouse made his air brakes famous with his work with the railroads and the Philadelphia Centennial Exhibition. In particular, the Pittsburgh, Ft. Wayne & Chicago Railroad teamed up with the Pennsylvania Railroads and lesser lines to link the West to the Exhibition. This line was known as the "Fort Wayne and Pennsylvania Route." The main route was put together to highlight the economical and safe travel by rails. The locomotives and cars, all equipped with air brakes, were used. The arrangement allowed for 145,000 passengers to move to Philadelphia daily from points west. Pullman Company "Hotel" cars lighted with gas were used. An excursion ticket from Chicago to Philadelphia was $32 for the round trip. Sleeping car arrangements added $10 and four meals an additional $6. Westinghouse rented out several trips to take all of his employees and families to the Exposition.

From 1870 to 1875, Westinghouse continued to maintain an experimental workshop, where he continued to amuse himself with his rotary steam engine models. During visits back to New England, George continued to work with steam engines as well. He applied for a number of patents on improvements of the basic steam engine. He often worked with his younger brother, Henry, on these new engineered controls. One of these inventions was the controller or governor for steam engines. This steam engine governor in 1875 became critical to improved steam ship control. This work with his brother and steam engines would ultimately lead to the formation of Westinghouse Machine Company.

In 1876 Westinghouse Air Brake was a world-class international business. The plant had 120 employees and a payroll of $75,000. The annual consumption at the plant was "900 tons of pig-iron, 200,000 bolts, 250,000 feet of pipe, 50,000 feet of rubber hose, 15 tons of rivets, 50 tons of copper, 20 tons of malleable fittings, 100 tons of merchant iron and forgings, and 37,500 bushels of coal."[16] The annual report noted that Westinghouse brakes were being used in twelve countries including Chili, Canada, Mexico, Cuba, Ecuador, Belgium, England, New Brunswick, Peru, and New Zealand, all from the

16 Samuel Durant, *History of Allegheny County* (Philadelphia: L. E. Harts, 1876), 120

Pittsburgh plant. Westinghouse was the first major exporter of industrial machinery. The Westinghouse exhibit at the Exposition was an extremely proud moment for Pittsburgh, remembering that Carnegie had just started his first Bessemer steel mill at Braddock a few months prior. Westinghouse was daily in touch with his European agents via telegraph. A new series of brake trials were being discussed for later 1877, which Westinghouse was preparing for with additional improvements.

Before leaving for Europe, however, Westinghouse would witness a major labor uprising at the doorsteps of his Liberty Avenue and 25th Street plant. The problem centered around the Pennsylvania Railroad with their 10% wage cut and implement of their "double-headers." The 'double-header" was the combining of two trains and two engines (two locomotives and 34 cars instead of one locomotive and 17 cars), but with a single crew. The rail workers were making a mere 5 to 6 dollars a week compared to Westinghouse Air Brake employees, which averaged $4 dollars a day. The workers of the Pennsylvania Railroad called for a strike on July 19, 1877, and spread from Maryland and West Virginia to Pittsburgh. The riot started on July 21, 1877 as the first "double-headers" arrived at Pittsburgh. Rioters overturned locomotives, pulled up tracks, and set the Union Depot on fire. 1384 freight cars, 104 locomotives, and 66 passenger cars were burned, eventually, over 3 million dollars of damage was done. Initially militia arrived on the governor's orders, and a battle ensued that left 20 dead. The battle inspired the mill workers of Pittsburgh's South Side who joined the fight on 28th street. The additional 1000 rioters pushed the militia back to the roundhouse on Liberty and 28th Street a block away from Westinghouse's plant. The troops mustered in an alley near the Westinghouse plant, which would become known as "Garrison Alley," the site of Westinghouse electrical research. On the 20th, the militia engaged again killing another 20 and wounding over a hundred. By July 24 citizen patrols helped to restore order. It was a struggle that further convinced Westinghouse of the importance of fair treatment.

World Fairs were sources of inspiration for people like Westinghouse. The Centennial Exposition was really the first to demonstrate the possibilities of electricity. Alexander Bell was there with his telephone, as well as a number of futuristic devices. Westinghouse returned to Pittsburgh to experiment with the use of electricity to activate the brakes. He worked up successful models, but found the wiring lacking the strength in actual operating conditions. This type of R&D work would become a hallmark of any Westinghouse organization. The Westinghouse system, however, was technically superior and Westinghouse was confident in the ultimate victory. With the support of the European journals and the Institution of Mechanical Engineers, a new series of trials was proposed for late 1877 into early 1878. These were known as the famous Galton-Westinghouse Experiments; Sir Douglas Galton was the British engineer in charge. Westinghouse returned again to Europe for an extended stay of almost two years. The Galton-Westinghouse experiments were true scientific trials with detailed measurements. The

Galton-Westinghouse trials went beyond the end results to study the science behind them. Westinghouse actually enjoyed the engineering approach of these trials and was very active during them. Westinghouse helped finance the needed testing equipment. The experiments generated many scientific papers looking at friction theory. The experiments ended in a victory for Westinghouse and a testimonial to Westinghouse the engineer and inventor. Sir Galton said the following of Westinghouse's development of the automatic brake, "The way in which Mr. Westinghouse had gone to work, directly he found that something was wanted, was as good an illustration of the spirit in which engineers ought to work as could be found anywhere." Westinghouse had become one of the most respected engineers in the new field of engineering. His methodology of observation and research, rough creation via stretching, then engineering drawing, followed by scale modeling, and finally scientific testing defines the discipline of engineering to this day. This pragmatic approach applied science to engineering. The title that is overlooked for Westinghouse is the father of industrial and manufacturing engineering.

Westinghouse and Douglas Galton took the data from the 1878-1879 trials and developed new engineering principles. Westinghouse and Galton studied the laws of physics throughout the trials. Some of their findings about friction changed the way engineers and scientists looked at this subject. These findings on friction were new and revolutionary. Galton and Westinghouse presented three papers on these findings before the British Institution of Mechanical Engineers. Mechanical engineers knew the practical "law" of railroad friction as the "Galton-Westinghouse law." Again, in these early trials we see the evolution of Westinghouse the engineer. These scientific studies on friction would ultimately lead to one of Westinghouse's greatest inventions — the friction draft-gear. Westinghouse had clearly evolved past the trial and error methods of many early Victorian inventors. He started to use science to narrow the scope of experiments needed. This is another example of Westinghouse's pioneering in the methodology of modern research and development. Men like Edison wasted endless hours in trial and error experiments, while Westinghouse eliminated many trials by the application of science.

Westinghouse was the philosophical leader of a group of Victorian inventors who defined the new field of engineering. These Victorian meta-inventors, such as James Nasmyth, Henry Bessemer, and Thomas Edison, applied science in new creative ways. The generations before them were true scientists, while they looked for applications before theory was fully understood. Their skills were their creativity. They were not specialists, but creators willing to chase economic needs and practical challenges in any scientific discipline. They believed that inventing was a profession itself. For example, Bessemer made millions from a stamp cancellation machine before he invented a new steelmaking process. Invention was seen as a craft, which would become the discipline of engineering. Westinghouse, more than any of the great Victorian inventers, pioneered the discipline of the engineering craft. His approach would evolve into the corporate approach to research and development used even today.

By the end of the decade, Westinghouse had arrived as a Pittsburgh industrialist. He had a membership in Pittsburgh's Duquesne Club, which from 1873 was the private men's club in Pittsburgh of some of America's greatest capitalists. His partner, Ralph Baggaley, had become of the Duquesne Club's earliest member and officer. For Westinghouse, membership in the club meant little; he preferred to eat dinner at home and would have nightly business dinners there as well. For lunches he preferred hearty basic meals and often ate at the Allegheny Valley Railway Company diner car known as "Coach No. 2." The bankers dominated the Duquesne club, which was another reason Westinghouse preferred doing business at home. Socially, Westinghouse preferred scientists and engineers to businessmen. He was also a member of the Shadyside Presbyterian Church, which was the church of prominent Pittsburgh industrialists. The Shadyside Presbyterian was the church of most Pittsburgh greats including Henry Clay Frick, Robert Pitcairn, Henry Laughlin, Mr. Dalzell, A. Childs, and many of Carnegie's lieutenants. He was a charter founder of Shadyside Academy, which George III would attend. Again, Westinghouse disliked these types of formal organizations. Westinghouse might well have been one of Pittsburgh's greatest Christians, but he rarely showed up at formal services. He fit the mold of the Pittsburgh industrialists of the time, Republican and Presbyterian. Westinghouse was a great capitalist as well, but he was never one of the "boys." He did not attend formal services at Shadyside often, nor did he hang out at the Duquesne Club. He and Marguerite where the earliest patrons of the Pittsburgh Opera (founded in 1871). He and Marguerite enjoyed Saturday and Sunday picnics with the families of his employees more than spending time with fellow industrialists.

CHAPTER 5. VICTORIAN ENGINEER

Westinghouse had a passion for science, engineering, and invention. He loved to take notes and sketch machines as he traveled, and found escape at his drafting boards. When in Europe, he visited factories and rode the rails to understand differences. He was a world traveler in time when travel was difficult. He almost always took Marguerite with him, so he would never suffer the homesickness of his early traveling days to sell frogs. The two years he spent in England were difficult because he was away from production and the factory, but the brake trials were a source of great satisfaction. Westinghouse built on his engineering skills, learning how to design and evaluate industrial trials. Time after time he turned trial failures into commercial successes. Even his competitors hailed his problem solving skills. Westinghouse's commitment to the European market resulted in seven years of effort between 1871 and 1883. He persisted in Europe against bias, and entrenched suppliers. By 1879 he had production operations in both London and Paris, as well as sales offices in Belgium, Italy, and Russia. While on the extended stay in Europe, he set up offices in England, France, and Russia, which would ultimately make him America's first international industrialist. This time in Europe would also introduce him to new areas of invention, such as signaling, railroad apparatus, and electricity. The brake tests had honed the engineering skills of Westinghouse as well. He, better than anyone, had come to fully understand the science behind railroading. Westinghouse realized better than anyone that improved brakes were only part of rail safety. The volume of traffic required that trains share tracks, thus head-on collisions were not uncommon. In the United States there was almost no administrative control of track switching and scheduling.

While British railroads lagged behind on brake systems because of their lighter trains, they excelled in the area of traffic control because of the population density. Traffic control affected railroads in several ways. First, there was the time issue. The same trip from Pittsburgh to Philadelphia could vary by as much as six to ten hours depending on "traffic," traffic relating to switching tracks and railroad crossing tie-ups. Another factor in traffic control was crossing accidents, such as the one his first air brake had avoided in September of 1868. Lastly, there was the issue of long stretches of shared track by trains moving towards each other. The cost of laying rail required this dangerous but necessary practice. Signaling and switching, therefore, went hand and hand. Signaling represented the art of moving trains faster versus stopping. The art of signaling was almost non-existent in the United States, but had early roots in Great Britain. The function of signals was to tell the locomotive engineer when to start, proceed safely, reduce speed, and when to stop. The reader needs to look at the technology of the 1870s to fully appreciate the problems facing the railroad in signaling. Electricity was in its infancy as was electrical lighting. Signals were usually audible such as whistles, or the visual flag type. If lights were used they were oil or gas. Train stations were connected by telegraph and would at least get an emergency communication to the train. The American art of signaling had not progressed past the 1840s "Delaware system." The station erected cedar poles with cheats so station men would climb up and with a spyglass check the status of the tracks. Flags and a hanging ball were then used to signal the train of problems or give the go ahead. The term "highballing" meant the ball was hanging high, which signaled the engineer to move a top speed.

England on the other hand, had developed block signaling in 1846 and interlocking signals by 1856. There were good ideas, but most inventors steered clear of signaling because there was little opportunity for profits. Westinghouse interest in safety and engineering system drew him in. Block signals hoped to maintain a minimum distance between trains. Using the block plan, sections or blocks were set every half to four miles. Each block had a station with a signalman at the beginning and end. As a train passed the station, a signal raised a danger flag to stop any other train from entering into the block. When the train passed the station, it telegraphed back to the previous station an "all clear," which took down the danger flag allowing trains to enter the block. It also allowed for safe track switching within a block. It worked well, but it was expensive. Fear of rail travel was causing problems on both continents and the Centennial Exposition at Philadelphia in 1876 was used to highlight rail safety. In the United States, the Pennsylvania Railroad was using the "Block Signal System" on a limited basis by 1876. The following is the Pennsylvania Railroad's Centennial Exposition review:

An important adjunct to the safe and expeditious running of trains is the Block Signal System. By this system the road is divided into sections between telegraph stations, technically known as blocks. The telegraph stations are ornamental towers, two stories high — the second story, which is the operating room being surrounded by windows, giving a clear outlook in all directions. The signals, so arranged that the engineer of an approaching train in either direction cannot fail to see them, are three in number; red being the danger signal, blue the caution signal, and white the safety signal. These signals are illuminated at night, and show the same colors as by day. A train approaches the station from either direction, and the engineer sees the white signal displayed. This indicates that the track before him to the next station, be it one mile or ten miles, is clear, and the train dashes on. Instantly the operator lets go the cord (for he is obliged to hold the red or danger signal out of view by hand) and the red disk is displayed again. Immediately on the train passing, the operator telegraphs the fact each way, and enters on a record sheet the train number and the exact time of its passing the station. The train having passed, the block it has left is clear, while that it has entered upon is closed. In a few minutes the click of the telegraph tells that it has passed the next station and that is also clear, and so on throughout the line.[17]

The lights were oil lamps with colored shutters, and short telegraphic code signals were used to simplify the spelling out of words.

It may sound like a foolproof system, but there were many accidents relating to human error of the operator, telegraph operator, and engineer. The advance of the interlocking system added a manual hand lever to the system. The hand lever controlled track switches and the corresponding signals. The "interlocking" was between the switches and the signal to prevent the wrong signal. Switching and interlocking signals were critical at major railroad terminals and yards. The hand British interlocking system had been shown at the Centennial Exposition in 1876 with few sales in the United States. Westinghouse had actually started researching signaling after his first trip to London in 1875. He had made drawings and taken notes to consider better designs. After his second trip, he returned to start making models.

The manpower need to maintain British interlocking systems was overwhelming. One British writer described the science at the London Bridge Switching Station:

Six hundred trains passed in a day and ninety trains in two busy hours. These figures do not include the many switching movements. This plant was worked by gangs of four hundred men in eight-hour shifts, and it took a husky man to pull over some of the levers.[18]

This amazing scene was repeated at stations all over Britain. British railways had tens of thousands of men employed in switching and signaling, and this amounted to a significant portion of the operating expenses. The government maintained a watchful eye to assure the safety of British railways. In the United States, big lines were hesitant to apply such manpower and cost.

---

17 James McCabe, *History of the Centennial of the Exhibition*, 1877
18 Prout, 218

The horrific headlines of head on crashes had started a political movement by the late 1870s.

The first improvement Westinghouse envisioned was the elimination of manpower to drive the interlocking switches. He logically thought of compressed air as the power source. Compressed air is a powerful and efficient means of transferring power, and even today remains used in factories for tough applications. Modeling again demonstrated that compressed air and underground lines could open and close switches while raising and lowering signals. This design had the same issues he had faced earlier with the straight air brake. Underground lines would require pumping stations to maintain pressure, and leaks would require digging up the lines to repair them. Again, Westinghouse looked at the possibility of using electricity to activate compressed air switches, yet allowing compressed air to do the work. The electrical current would eliminate the need for compressed air lines, the electrical signal could activate a local reservoir of compressed air that would power the switch and signal. It was the same idea he had tried with air brakes, but there the wear and stress of coupled cars broke the electrical wires readily. This of course would not be a problem with track switches. This air-electrical combination, Westinghouse dubbed as "electro-pneumatic" control. Switching and signaling problems were an engineer's nightmare requiring the best science available, applied mathematics, and experimentation. Maybe the most important result was Westinghouse that started a personal study of the status of electrical engineering. Westinghouse did not really enjoy reading other than the daily newspaper, but he had learned back in New England the need to focus his library research prior to launching experimentation.

After his return from Europe in 1878, he increased his experimentation at the Garrison Alley plant. He came up with the idea of using the metal track to carry the electrical current. Personally, he was now extensively researching journals and books on electricity, which would pay dividends later. Using the track as the electrical conduit was simple but powerful. First, it reduced the expense of copper wire, which was extremely expensive in 1878, and second, a track break would stop the current and signal a problem. Westinghouse devised a signal that would be activated by a track break. Furthermore, he devised a system to automatically activate the brakes on a train in case the engineer missed the danger signal. He was concerned about the need for pumping stations, reservoirs, and freezing of compressed air switches and started to work with hydraulic fluid power in place of compressed air. Westinghouse experiments concerned Baggaley and the other directors because they saw limited commercial value, but Westinghouse was a visionary running about ten years ahead of the market. Westinghouse was convinced that air brakes coupled with automatic interlocking signaling and switching could eliminate railroad accidents. The directors also did not want to dilute Westinghouse's efforts; air brakes were paying the bills. Garrison Alley was quickly becoming a "Menlo Park" laboratory for Westinghouse. While Westinghouse was approaching full penetration of the passenger train air brake market, he had converted less than 5% freight trains, and there were 40 freight cars for every passenger car. While his own directors viewed his experiments in signaling as a wasted time on "elaborate frills," a superintendent of a small railroad noted, "If the men who work the railroads ever choose a patron saint it will be St. George — in honor of George Westinghouse."

Undaunted by his own board of directors, Westinghouse ploughed forward with experiments and modeling. In 1880 he had enough data and a good design for detailed engineering drawings, and he filed for several related patents. All Westinghouse's ideas were not totally new; a company in Massachusetts was producing electric signal devices, which Westinghouse was also testing in Pittsburgh. This Massachusetts Company seems to have been the source of his idea to use the track as an electrical conductor. Another company in Harrisburg, Pennsylvania, Interlocking Switch and Signal Company had some key patents on switching devices. Westinghouse had actually started buying stock in these companies in the late 1870s. Westinghouse also purchased patents if necessary. He originally had hoped to bring Air Brake into the field of switching and signaling, but his directors opposed any such move. Westinghouse was never conservative about money when it came to something he was passionate about. He used all his personal funds including his house as collateral to buy controlling interest in Interlocking Switch and Signal Company. He could have moved into the field without buying this company, but he wanted the key switching patents they owned. Again, we see the character of Westinghouse in his respect for intellectual property. Using these patents, he formed a new company — Union Switch and Signal. The new company started as a small machine operation on Garrison Alley, Pittsburgh. It was a necessary step, since the board of Air Brake felt they could not support this engineering incursion into a new field. Westinghouse, however, did bring in some European engineers such as Guido Pantaleoni and Albert Schmid. Pantaleoni was the son of an Italian physician, Doctor Diomede Pantaleoni, who treated Mrs. Westinghouse when she suffered from a deadly illness on the 1878 European trip. Pantaleoni was young, but Westinghouse deemed him competent. He was also a graduate of the University of Turin. Pantaleoni also had interested Westinghouse in a process of producing artificial marble from gypsum! Schmid was already a well-known Swiss engineer, especially in the application of arc lighting. Westinghouse had met him initially at the Paris Exposition. Union Switch and Signal would become an incubator of ideas, which would ultimately lead to Westinghouse Electric Company.

His endless experiments on switching and signaling lead him into another area in the late 1870s. Westinghouse was one of the first to get a telephone in Pittsburgh, as we were involved in his signaling experiments. The telephone was limited to about ten miles at the time and required a direct wire-to-wire connection. Westinghouse had found out in his electro-pneumatic inventions, as had Edison, that copper wire was a major constraint because of its high cost and short availability. The earliest adopters of the telephone included the Pittsburgh banks and the great industrialist homes of Homewood. If Westinghouse wanted to have a telephone connection with ten banks, he would need ten wires to each of banks. Party or shared lines in neighborhoods helped save wire. Westinghouse envisioned an exchange or switching station, which like railroads would allow shared lines by automatic blocking. He applied for his automatic telephone-switching patent in October 1879. He ultimately applied for and received patents based on his detailed system drawings. He was actually many years ahead of Bell Telephone, but he lacked the time needed to fully pursue its commercial application. While operator switching had started at Bell Telephone, it wasn't until 1920 that automatic machine switching came into use. Only Thomas Edison could match the diversity that Westinghouse demonstrated in creative inventions.

Westinghouse falls into a unique group of Victorians known as meta-inventors.[19] Meta-inventors had great cognitive skills and problem solving skills. They were generalists who saw inventing as a craft. They included Bessemer, Bell, Faraday, Ford, Morse, Nasmyth, Edison, and many others. They rarely advanced science, which was the work of specialists, but could bring streams of new scientific discoveries into practical devices and systems. Westinghouse exemplified the use of a problem as the motivation to invent something. They were problem solvers more than engineers or scientists. Their cognitive skills bridged the left and the right brain. They would deduce general principles from systems and apply them in new areas. Westinghouse could see the same system concepts in railroad switching and telephone switching. Later he applied similarities in compressed air delivery systems to that of natural gas delivery systems. Meta-inventors could also apply their cognitive skills to a problem representing the most economic value. They are rare today because science and engineering have become very specialized. The Victorian generalist or "man for all seasons" had little difficulty breaking the great barriers of entry put up by specialists. Westinghouse cared little for such barriers, and in many ways he was equally belittled as an engineer and scientist.

In 1880 he started Westinghouse Machine Company to build a type of high-speed steam engine developed by his younger brother Herman. These engines would drive electrical generating dynamos for arc lighting and by 1882 Westinghouse Machine was producing 4 engines a day. It seemed to fit well with his small research and development operation at Garrison Alley, which was pursuing electrical switching controls. The actual manufacturing plant for Westinghouse Machine was the original Air Brake plant at Liberty Avenue as Air Brake moved to Robinson Street on Pittsburgh's North Side (then called the City of Allegheny). His new Air Brake plant would be world class. Westinghouse had molding machines imported from Scotland when

19 Quentin R. Skrabec Jr., *The Metallurgic Age* (Jefferson: McFarland, 2005)

he couldn't find the automation he wanted domestically. His lathes were some of the first gear and cam driven lathes in the area. Westinghouse studied tooling with a passion allowing him to always be on the cutting edge. He also got advice from his father and brothers who still ran a machine shop in New England. The workers appreciated his smaller design points as well, which included locker rooms, washrooms, and showers. The workers enjoyed using their Saturday afternoons off to play baseball at near-by fields, which Westinghouse helped to maintain.

Westinghouse was now realizing that his creative ideas were beyond his ability to personally develop them, a realization the Air Brake directors had come to years earlier. He started a number of electrical lighting experiments in the early 1880s at Garrison Alley. He was evolving into a manager of inventors, projects, and companies. It was a transition that Edison would never make. Edison's Menlo Park was also a great creative incubator, but Edison remained the head inventor, and like a master craftsman, remained dominant in the process of inventing. Westinghouse was able to set the direction and monitor progress while allowing his employees to become inventors of their own. It took humility and a respect for intellectual property for Westinghouse to function in this new role, but it was one he was most suited for. Westinghouse, of course, would always have his pet projects. From 1880 to 1883, Herman and George were still working on various steam engines. They started to produce a small marine steam engine for small yacht. The 35-foot yacht, the Topsy, was able to achieve speeds of 15 miles per hour with the Westinghouse engine.[20] The engines were also gaining wide use in factories for driving main machine shafts. Mechanics hailed the Westinghouse engine for its tight tolerances, bearings, low vibration, and offset drive. They were also "self-lubricating" requiring almost no daily maintenance. To achieve this self-lubricating feature, a new bronze-bearing alloy was developed. The Westinghouse engine became the darling of industry in a few short years. By the 1890s Westinghouse Machine was producing 60 steam engines of various sizes. Just as profitable was his construction firm Westinghouse, Church, Kerr, and Company that applied these engines to factories. These engines were unheralded monuments to the genius of both George and Herman Westinghouse. Westinghouse also worked on agricultural equipment for his father's firm of Westinghouse Company back in New England.

Westinghouse engines were some of the best built in the world. The engines profited by the pioneering use of drafting machines and machine parts. The Westinghouse approach forced the designer to consider the fit and tolerances of the parts. This may seem only logical, but it was a time when machinery was built by blacksmiths and small machine shops. Drafting or parts drawing forced a standardization of the parts and supplier processes. Drafting was as revolutionary in the 1870s, as E-building and desktop manufacturing is today. It prevented costly trial and error fitting on the ma-

20 "American Institute Fair," *Manufacturer and Builder, October, 1883*

chine shop floor. His machines and equipment were renowned for their tight tolerances. This approach allowed Westinghouse to take the crafts based American machine into the Industrial Revolution. As we have seen, the roots of this evolution go back to his father's machine, which functioned in the transitional period. It was Westinghouse that was idolized by engineers, not Edison. Westinghouse did more than anyone previously to make the field of engineering, a true profession.

Meantime, by 1881, Westinghouse Air Brake dominated the world in air brake production. He basically owned the brake market for passenger trains the world over. In the United States, he had outfitted 3,435 locomotives and 12,790 cars; in England it was 1,087 locomotives and 7,719 cars; and in France, it was 1416 locomotives and 7,193 cars. The many years in Europe had paid off. Westinghouse at manufacturing plants in England, France, Germany, Belgium, and Russia. His world capacity was the ability to equip three hundred locomotives and twelve hundred cars a month. Westinghouse Air Brake was the largest international manufacturing enterprise in the world. Only one other company had had such exporting success, and it was the Pittsburgh Company of H. J. Heinz, whose products the English had developed a taste for. Air Brake had now made Westinghouse a multi-millionaire, and probably at the time the richest man in Pittsburgh, outside a few Scotch-Irish bankers and railroad owners.

Providence would change his direction in 1884. Marguerite had become pregnant after years of trying. They were particularly concerned because of Marguerite's propensity to illnesses. As the birth neared in the winter of 1882, Westinghouse brought in a New York specialist. Finally, they left for New York to have the baby in late December. Westinghouse would not be separated from his wife and rented a suite of hotel rooms for business and living. After months of worry, the birth was perfectly normal. George Westinghouse the third was born shortly into 1883 during a sub-zero cold spell. The Westinghouses decided to stay in New York rather than travel in winter with the newborn. They would remain there with George at their side. Family always had priority for Westinghouse even if he fit the description of a workaholic. He had his brother Herman watching over Garrison Alley, and Air Brake was self-sufficient. As was his practice, Westinghouse had the local Pittsburgh papers delivered to him daily via train. Westinghouse loved to start his day with the morning newspaper. It was in the local press that he read of a natural gas find in Murrayville Pennsylvania east of Pittsburgh. Natural gas offered potential as a future fuel if a distribution system could be devised. He had hired a "grimfaced, middle-aged nurse" who limited access to his newborn son. Even Marguerite was forced out and resorted to shopping in New York. Still, Westinghouse would not leave New York until he could bring little George back home. Westinghouse, having some time to spare used the extensive libraries of New York to research the subject.

Gas distribution was another area he had become familiar with in Britain. Coal gas had been used to light major cities of America since the 1830s, but it had not become popular as an illuminant in homes after the Civil War. America tended to use kerosene lamps, which produced a smoky residue through the house. Gas offered a clean burning alternative to kerosene. Coal gas was the product of burning and processing coal in retort furnaces. Coal gas had started to become popular at the end of the 1870s, just as Edison invented the incandescent light and arc lighting was becoming available for city streets. Edison's work had sent gas company stocks tumbling. In England, Westinghouse had seen the huge potential market for coal gas home ovens that was non-existent in the United States. Natural gas, in addition, had found few applications because it was limited to the well area lacking any type of distribution system. Also the abundance of coal in the Pittsburgh area had made any other fuels too costly. However, in 1880 Pittsburgh was the glass capital of America, which also needed a cleaner fuel. These glassworks were a natural for the Pittsburgh area with its river sands and cheap coal as fuel. It was an extremely dirty fuel and this was problematic for glassmakers. Rochester Glassworks near Pittsburgh had been experimenting with natural gas and Westinghouse had followed their progress.

Several small natural gas companies were struggling in the Pittsburgh area in 1883, and were plagued by many problems. Supply problems to factories caused by pipe leaks were common. Home connections were even more problematic. Cock valves were used to shut the gas on and off. Natural gas is odorless and if cocks were not shut, rooms were filled with asphyxiating gas. Explosions were also common and there were even local laws restricting gas usage. If this wasn't enough, these gas companies had no way to charge for usage, so lacking meters you were charged for a connection regardless of use. The earliest uses of natural gas were by companies such as the iron works of Spang, Chalfant & Company, which had a gas well on site. Natural gas was found in1882 in Murrayville, east of Pittsburgh. The Murrayville find suggested that natural gas was more abundant than originally thought. Westinghouse even started to study geology and believed the soil around Solitude suggested gas deposits. In reality, there was more wishful thinking behind Westinghouse's soil samples than science.

They returned to Pittsburgh in March, and Westinghouse hired geologists to explore the grounds at Solitude near Thomas Boulevard. By the end of 1883, Westinghouse had started to drill on his property. He beamed with enthusiasm and spent nights designing drilling tools to amuse himself. He encouraged the drilling crew and spent hours talking and taking notes on drilling. One night the well came in with a major explosion and a hurricane like hiss disturbed the once peaceful residential neighborhood. To test the well, it was lit, producing a huge column of bright light that could be seen for many miles. The brightness was described as The gas lamps of the city dwindled to a little points of light, and persons in the street not less than a mile away were able to read distinctly the finest newspaper print by the light of gigantic natural flambeau on the heights of Solitude." The Pittsburgh fire department had to be summoned to hose down the mansion to prevent it from burning. The tremendous pressure at the wellhead was obvious, and the first step was to design and make a stopcock to cap the well. Westinghouse handled this task personally as the neighborhood awaited some relief.

The problem how became now to get into the business of distributing natural gas. Political opposition was rapid headed by the Fuel Gas Company, which had a public utility charter for the Pittsburgh area. The opposition was extremely stiff as the Fuel Gas Company formed a confederation including the coal gas companies and some banks to oppose him on all fronts. Westinghouse's personal lawyer and future congressman, however, were up to the task. John Dalzell found a utility company charter in Harrisburg that could be purchased. This charter for "Philadelphia Company" had been set up by Thomas Scott of the Pennsylvania Railroad for railroad lines but had an open ended charter that could be used for any utility. Westinghouse was able to purchase the charter for $35,000. On August 4, 1884, the Philadelphia Company was formed with George Jr. as President, Robert Pitcairn as vice-president, and John Caldwell as secretary. Herman Westinghouse and John Dalzell were also officers. He needed a great deal of stock subscriptions as well. The banks did not help much and Westinghouse sold most of the stock to personal friend, well-known Pittsburghers, such as A. M. Byers, C. Jackson, and John McGilley.

The next problem was a fight with the City of Pittsburgh to lay pipe for gas distribution. The politics were problematic with competing fuel companies arguing that he would have an unfair monopoly. Citizen groups argued the danger of major explosions. Westinghouse was, however, very focused on the engineering side of the project, not the potential financial gains. He laid out all his safety measures to the press. At Westinghouse Machine, he designed new pipe connections to prevent leaks and assure safety. In addition, he offered free gas to the city's fire stations and police stations. He offered to cooperate with the few small gas companies allowing them to use the lines at a nominal charge. The city council was won over by the fairness of Westinghouse in these business offerings. The city also realized that cheap fuel could rejuvenate industry growth that had slowed from the Panic of 1873. The glass industry was also under pressure as gas finds in more western states were creating a loss of glass houses. Even Pittsburgh's steel and iron industry's growth had slowed.

Westinghouse's Philadelphia Company was profitable within the year. Westinghouse had applied for a patent for his distribution system. By the end of 1885, Westinghouse had applied for twenty-eight patents in the gas distribution field. The pressure presented a major problem and had become the talk of the town with a major accident. A man had struck a match only to set off a major explosion throwing him and a number of horses thirty feet into the air and starting a major fire in the city. There were even cries to ban the use of natural gas. Westinghouse was well prepared to deal with high-pressure gas having worked with steam and compressed air most of his life. In fact, few people in the world had the kind of experience that was needed to solve the problems of natural distribution. The pressing concern to calm the public was to eliminate leaks. Westinghouse again applied the simple principles of redundancy he had used in braking hoses and pipes. In the case of natural gas, he used a double pipe that is one inside another. The leak in the inner pipe would be captured by the outer pipe and the pressure was diffused. He then devised a system of escape pipes along the street mains to feed leaking gas into corner lampposts or to escape at safe points to the air. High well pressure was great for transmission, since pressure pipe diameters using the smaller diameter and high pressure for transmission then at the point of use provided force to move gas, but it made gas unmanageable at the point of delivery. He designed a series of increasing diameters along the line until the point of delivery, which had the bigger diameter to reduce pressure, the reduced pressure making it more manageable at the point of delivery. For example, the first four to five miles of pipe was small diameter to speed gas into the long distribution pipeline, then every five miles or so the diameter was increased. Using various diameters from 8 to 36 inches, Westinghouse moved gas up to 20 miles from the well by changing pipe diameters; the pressure was reduced from 300 pounds per square inch to a few ounces per square inch. Later the Philadelphia Company added pumping stations for even more distance. This simple, yet brilliant engineering system would be used later in his career to solve the problem of AC electrical current distribution.

Gas heating, cooking, and lighting became very popular in Pittsburgh, but not without some hesitation by the public. The very fashionable and earliest of the Pittsburgh's suburbs to convert to gas, Wilkinsburg reported the following:

Using gas was a new experience to people and most of them were afraid of it, and not know whether to turn on the gas first and then light it, or just how it should be handled. Many singed eyebrows and burned hands resulted, and even some bad burns. The gas pipes were run to the second floors on the outside of the houses. In grates and stoves T burners were used with broken pieces of firebrick. A can of water was usually set on the stove or hung before grates on account of the dry heat. Gas was plentiful and a six-room house could be heated for about $2 a month [in a time with the typical weekly pay was around $10].[21]

From 1883 to 1886, gas distribution was his passion as he advanced natural gas distribution to a science. The growth of Philadelphia Company was amazing having 45 wells and 330 miles of pipe by 1885. Lines reached far beyond Pittsburgh to Homestead, Braddock, and McKeesport. One safety problem was household explosions, which had been a major deterrent to its use. These explosions resulted because of frequent shutoffs to a house, while the gas flowed to the lights when on unless the household shut the incoming valve. When the distribution line was disrupted and undetected by the household, the light went out and upon its return odorless gas filled the

21 Harry Gilchrist, History of Wilkinsburg Pennsylvania (Pittsburgh, 1940), 104

house. Westinghouse invented a safety device known as the automatic cut-off regulator. When the gas pressure held below the four-ounce household working pressure, the gas was automatically shut off. Westinghouse started to convert his Homewood neighbors, such as Henry Clay Frick's Clayton. From the management of the company, Westinghouse invented meters for household and factory use. The company could now charge by usage. Westinghouse continued to improve the system with the invention of leak detectors and improved drilling techniques. He also invented air-fuel mixing valves to make industrial furnaces efficient. Eventually, Westinghouse was granted thirty-eight patents in the natural gas area. Many of these greatly improved the wasteful distribution of natural gas. Prior to Westinghouse's system, natural gas was vented and lighted to burn every weekend. In particular, downtown Pittsburgh was lighted by a battery of stand pipes on the hill over Union Station from Saturday to Monday to burn off surplus gas from the factories shutting down on weekends. It was even said that for a few years Pittsburgh's skies cleared with the use of natural gas, only to again blacken with the exponential growth of the iron and steel industry. It was estimated that 1000 cubic feet of this gas was equivalent to 56 pounds of Pittsburgh. The cost of 1000 cubic feet of gas was only 10 cents. In 1887 it was estimated that natural gas had replaced 40,000 tons of coal daily in the Pittsburgh area.[22] Westinghouse's gas industry was starting to seriously impact the coal business of his neighbor Henry Clay Frick; even though Frick had been one of the first to light his home with gas. The price of coal had fallen 35% to 40% by 1887 and estimates of 10,000 miners were being put out of work by conversion to gas.[23]

The amazing change in the Pittsburgh environment in 1889 was described as follows:

> The transition from the use of coal to that of this non-producer of smoke has been simply wonderful. Prior to 1884 the city used 3,000,000 tons, or nearly 80,000,000 bushels, of bituminous coal in a year, the smoke from which hung in a black cloud, like a pall, over the city continually, discharging flakes of soot and fine dust in a steady downfall. The black, pall-like cloud has disappeared; the atmosphere is cleaner than that of most western cities, and the temptation to use bright colors in house-ornamentation is steadily growing. The consumption of coal has fallen much below a million of tons annually.[24]

The return of the heavy smoke and dirt in the 1890s, most certainly, was disappointing to Mrs. Westinghouse, who had always struggled with this feature of Pittsburgh living. To some degree the return of the black skies was attributable to Westinghouse's own success. Westinghouse's system and his Philadelphia Company were extremely successful at converting homeowners to the use of gas, but many factories concerned over supply moved back to coal. Another interesting benefit of gas home heating was related in

22 *Handbook of Pittsburgh and Allegheny* (Pittsburgh: Fisher & Stewart, 1887), 26
23 *Manufacturer and Builder*, January, 1886
24 *History of Allegheny* (Chicago: A. Warner & Co., 1889), 616

the press, "The use of gas allows people who have to rise early, at least half an hour more sleep owning to the fact that there is no delay in lighting fires. The Philadelphia Company thus contributed 5000 extra hours of sleep to its customers."[25] Westinghouse even solved the initial lower candlepower of the natural gas compared to coal gas by adding moisture. By 1888 Philadelphia Company was supplying over 25,000 homes as well 700 factories.

The savings of gas cooking over coal stoves was also substantial. It was assumed the average cost of $12.50 a year; others stated that it took two pounds of coal to cook a pound of meat. Westinghouse had contracted several professors to calculate the savings from gas. It was estimated that for the same amount of coal, a family would use 3000 cubic feet of gas (at 50 cents a thousand in Boston) at a cost of less than $2 a year.[26] Another small, but for cooks, a revolutionary device of Westinghouse was a temperature control valve in 1887. England was far ahead of the United States in gas cooking, but areas like Pittsburgh were moving forward. The problem became that many areas experienced gas booms and busts in a period of years, creating fears about the supply. Toledo, Findlay, and Bowling Green experienced gas booms lasting only a few years. Geologists were hard pressed to make accurate estimates. Still in areas such as Pittsburgh and West Virginia, the supply seemed endless.

The following poem about natural gas appeared in the Pittsburgh papers:

THE BLESSING OF NATURAL GAS

Oh, this natural gas is a wonderful thing

And it giveth to dallying blessings a wing

And to many a sigh

It does give strength to fly

And it maketh the lazy man merrily sing;

When he comes home at night there's no kindling to chop,

There are no lumps of coal on the carpet to drop,

There's no hatchet to find,

And no ashes to blind,

And there's no pesky grate to go flipperty-flop.

Ah, the hours of the night he can happily pass,

He may dream of the summer, the flowers and grass,

There is no fire to build

For, to his task skilled,

25 *Pittsburgh and Pittsburgh Spirit* (Pittsburgh: Robert Forsythe Company, 1928), 129

26 "Cheap Gas for Heat and Power Purposes," *Scientific American*, April 18, 1896

He will lie in his bed and turn on the gas.

Westinghouse was to play a key but unheralded role in making Pittsburgh the steel city. Westinghouse as a railroad engineer had been one of the biggest promoters of using steel in railroad application. By 1875 Carnegie had opened his massive mill in Braddock to make steel railroad rails, and by 1888 Carnegie had expanded to another mill in Homestead. Carnegie competitors were also locating in Pittsburgh. The emerging steel industry used coal to fuel its iron making, but needed gas to heat its many reheat and steel working furnaces. The rise of the new steel technology of open-hearth steelmaking would require gaseous fuel. Carnegie's Homestead mill would be the largest open-hearth steel mill in the world and demanded massive amounts of gaseous fuel. A cheap supply of natural gas made Pittsburgh the logical choice for the heart of the industry. It also helped save a large portion of its glass industry, which was now dependent on a cheap supply of natural gas and was moving west. Natural gas offered significant cost reductions to manufacturers in all industries. Westinghouse's entrance into gas distribution saved many hundreds of high paying glassmaking jobs for Pittsburgh. It is clear that Westinghouse more so then Carnegie made Pittsburgh a industrial center, but he would be doomed to publicly follow in Carnegie's shadow. Still Philadelphia Company made huge profits for Westinghouse and its investors. It also helped build a major nine-story office building in downtown Pittsburgh at Ninth Street and Penn Avenue in1888, which overlooked the site of his first air brake trials. By 1890 Philadelphia Company was the largest gas company in the world, having 900 miles of pipe.

By the end of the 1880s, Westinghouse and Philadelphia Company's gas distribution system was being hailed as a great feat of engineering. Writer George Thurston noted in 1888, "Mr. Westinghouse's achievement is one at which the wonder grows as time elapses, and the full results are realized. The transformation in all things, whether in the factory or the household, was almost like a page out of the Arabian Nights, and to the people of Pittsburgh it is still a wonder, and almost unrealizable."[27] In a few years, Westinghouse had changed the nature of energy usage in Pittsburgh and the country. The engineering devices needed were invented in a couple of years. He organized an army of 5000 plus men to lay pipe and another 1000 in the drilling of wells. He addressed every problem with a determined application of engineering principles. As he had with rail travel, he made gas heating and lighting a safe operation. He created well over twenty thousand related jobs for the Pittsburgh area.

Records of Pittsburgh show that in 1887 nine-tenths of the factories and three-fourths of the homes were using natural gas. Still investors in the mid 1880s were becoming nervous as Edison's incandescent lighting was gaining popularity and there was concern about the natural gas reserves. The

27 George Thurston, *Allegheny County's Hundred Years* (Pittsburgh: A. Anderson & Son, 1888), 205

geological and scientific understanding of natural gas was lacking. Reverses could not be estimated, leading to uncertainty of supply. Some wells ran out in a few years, while others, such as some in China are thousands of years old. In 1887 Westinghouse formed the Fuel Gas and Electric Engineering Company to experiment with the production of fuel gas from coal. He envisioned gas turbines for generating electricity more efficiently than steam. The experiments of making gas from Pennsylvania coal were costly and ultimately failed, but Philadelphia Company was generating plenty of cash for the engineer in Westinghouse. Westinghouse was also becoming interested in his electrical experiments at the time. The successes of the air brake, machine, gas, electric and others required him to build a corporate headquarters for all of these in Pittsburgh. The building was somewhat of an engineering triumph of its own. It was a 20 story steel building. It was heated by gas indirectly using steam radiators and lighted with incandescent lights. It also had three high-speed passenger elevators, all driven by a Westinghouse steam engine. His 20 stories would top that of the 14-story Carnegie Building and be dominating in Pittsburgh until Frick built a 22-story office building in Pittsburgh to claim the title.

By the end of the 1880s, Westinghouse was one of the famous Pittsburgh millionaires, with neighbors such as Andrew Carnegie, Henry Clay Frick, and the Mellons. Mrs. Westinghouse was followed by all the social columns of the day. An 1888 review of Pittsburgh's wealthiest women noted:

Mrs. George Westinghouse, Jr., of Homewood, lives in greater style, entertains more splendidly and wears more gorgeous, varied, elegant toilets, has more and finer diamonds than any woman in Pittsburg. Her table appointments are simply superb, the entire service being of solid silver and gold (whose cost it would be idle to attempt to guess), and the cut glass, Sevres, Dresden and other fine porcelains are worth a small fortune. Their whole style of living is after the plan of the household of an English lord. The house is a perfect palace, and the grounds worthy of it. Their stud of horses comprises some magnificent animals. Mr. Westinghouse is blessed with the knowledge that he was married for love, and not for lucre — something rich men cannot usually congratulate themselves on, as when he won his wife both he and she were but moderately well off. Mrs. Westinghouse is notable for splendid charity, both public and private. Her husband's wealth runs well up into the millions, $5,000,000 being an underestimate. Mr. Westinghouse was knighted a few years ago by King Leopold of Belgium as a token of admiration of the master mind that conceived the Westinghouse Air Brake. The brilliant young inventor and his wife would therefore be entitled, did they wish, to style themselves Sir George and Lady Westinghouse.[28]

28 Adelaide Mellier Nevin, *Social Mirror: Character Sketch of the Women of Pittsburgh* (Pittsburgh: T.W. Nevin, 1888), 94

Chapter 6. American and International Industrialist

The rise of the Philadelphia Company and its success was far more than a triumph of engineering; it was the perfect business venture. Westinghouse had managed to overcome insurmountable political and government regulations, he worked cooperatively with his competitors for an agreement, and financed the project without the help of the Scotch-Irish master bankers. It was a triumph that the bankers would remember. Westinghouse was gaining a national reputation in many fields. There was no mechanical engineer his equal. By the late 1880s Philadelphia Company was extremely profitable and Air Brake dominated its market. Even his Union Switch and Signal was growing and required a new plant. In 1886 he moved Union Switch and Signal to the borough of Swissvale. Swissvale was a rural suburb just a few miles from his Solitude mansion. His Garrison Alley operation was a world-class research center rivaling Edison's Menlo Park, and in 1886 it evolved into Westinghouse Electric. His machine company, mainly under the management of his brother, was making high steam engines at the rate of 60 a month, and was profitable. Still, there were mounting challenges especially in the air brake business. Freight trains in the United States had held out on conversion and Europe still clung to the use of vacuum brakes. The whole field of electrical power and the application of electrical devices had also got his interest.

The decade of the 1880s provided some of the happiest times for the Westinghouses. The businesses were booming and his biggest battles were still a few years away. While George and Marguerite were very social people, they did not move in the city's high society normally. Marguerite was a charming and beautiful wife, and like George, she loved to entertain quests at their home, Solitude. Westinghouse remained unusual in that time for business done over dinner at the Duquesne Club or Pittsburgh Club, Westinghouse preferred to bring businessmen to his home with their wives for dinner. Marguerite seemed to love home entertainment and these dinners became a daily thing when George was in town. They often mixed local Pittsburghers with famous scientists and foreign royalty. Talk would often center on their growing collection of European souvenirs. Marguerite herself loved European fashions and was known throughout Pittsburgh for her fine clothes, but she was also fond of the fashions found in Pittsburgh stores such as Kaufmanns. George also was fond of shopping in Pittsburgh for his long overcoats and top hats.

While Marguerite loved white she was restricted to black because of endless dust and smoke of the city. George was known to prepare salad dressings, even thro they had a full staff of cooks, maids, and butlers. The Westinghouses loved people and the staff was already treated like part of the family. Often these famous visitors left the special train crew waiting for the end of the party. Westinghouse would often invite the crew to join them for dinner. Their black butler Isaac Watson was a good friend as well, and was always ready to help Marguerite with her gardening, and George with his late night designs. Westinghouse had developed a somewhat unusual habit for a man that loved drawing; he seemed to have never carried a pencil. In his factories, employees were used to "lending" him a pencil at his call. At home it was Isaac Watson who would stand ready to have a pencil for Westinghouse. This habit of lacking a pencil seems to standout compared to other engineers of the time who seemed obsessed with their pencils. Edison, for example, had them made special in almost stubs for his use.

The decade of the 80s included the continuing battle for the European air brake market and the American freight train market. The air brake market was also becoming competitive again by 1885. The vacuum brake saw an increase in European sales, which had always had popularity in Europe; the vacuum brake supplied a slow but gentle stop. In Europe freight trains were shorter and lighter so the vacuum brake made sense. Two newer types of brakes, the friction and buffer, were starting to gain some market share. The mixing of different brake types on cars presented an unacceptable situation for the railroads overall. Free interchange of cars among different railroads was critical to the smooth operation of the railroads. Cars of the Pennsylvania Railroad, for example, needed to have the ability to be picked up by the Union Pacific or the Atchison, Topeka and Santa Fe Railroad. The same brake system was needed to facilitate this. You could not mix vacuum and air brakes, for example. The railroad companies, therefore, wanted a winner take all decision. The representative group of the railroads, the Master Car Builder's Association, started to call for standardization in 1884. Government regulators started to take notice as well. In particular, a Mr. L. Coffin, State Railroad Commissioner of Iowa, called for standardization. Westinghouse saw standardization as necessary as well, and he had always applied it in his manufacturing methods.

This would lead to the Burlington trials for freight car braking in1886. The Chicago, Burlington, and Quincy Railroad at Burlington, Iowa arranged these. The Master Car Builder's Association would act as judge. Many Air Brake managers were concerned with the "winner take all" atmosphere of the trials. George Westinghouse, however, loved the idea of scientific and engineering trials. He had learned a lot from his participation in the British trials and was confident. All major brake manufacturers were invited to participate in trials of fifty car trains. The initial trials in the spring of 1886, basically eliminated all types except vacuum and air brakes. However, no system met the standards needed for freight emergencies. Application of the brakes suddenly resulted in uneven stopping. Front cars stopped immediately, but the balance swagged and banged. Animals such as cattle would be hurt or killed in such instances. The judging committee called for further trials the following year.

Westinghouse was a tough competitor because he poured over the trial data and studied every problem, while living in his "Pullman Palace." Within weeks he was designing and modeling modifications to his brake systems. He convinced the Air Brake directors to spend two hundred thousand dollars to fit a fifty-car freight train of the Pennsylvania Railroad. Even with Westinghouse Air Brake booming profits this was a huge amount for a demonstration, but Westinghouse felt it was critical to win the freight car market, which was wide open and huge. Westinghouse started to make improvements in several areas including his triple valve device and the use of electricity to activate the compressed air. His improved triple valve conveyed braking force to the last car in the fifty car train nearly twice as fast as in the 1886 trials. Still, he could not dramatically reduce the violent shocks of sudden stops. The 1887 initial trials had very mixed reviews. The biggest surprise was that his automatic air brakes lost to the vacuum brakes. His application of experimental electro-pneumatic brakes performed satisfactorily. Westinghouse, however, had no confidence in his own electrical activated brakes. He believed from earlier studies that wires were too easily broken, which resulted in no braking. Westinghouse also was in no position to change over to these new experimental electro-pneumatic brakes. Some of the managers of the trials recalled that Westinghouse started drawing new designs as he watched each failure. At Westinghouse's 90th Anniversary, W. Nichols recalled Westinghouse's design breakthrough at the trials; "Finally, one morning, he [Westinghouse] left impetuously before noon, inviting me to join him at lunch in his private car on the West Burlington Shops siding. Later, I passed through the corridor to the dining room, he emerged from his stateroom and waved a bit of brown paper, exclaiming with characteristic enthusiasm, 'Here the triple [valve] that will revolutionize railroading!' How true that proved."

Many in the industry thought the initial trial results doomed the automatic air brake. Within months of Westinghouse's fall from engineering supremacy, Westinghouse redesigned and applied for a patent on his "quick action triple valve," he had drawn on that brown piece of paper. The final trials had proven Westinghouse's design. This valve met and exceeded every criticism leveled at his brakes during the 1887 trials. It became know as the "Quick acting Triple valve," and it went on as he had predicted to revolutionize the railroad industry. It would allow a fail-safe application of air brakes on long freight trains. With the trials over and the word spreading throughout the industry of Westinghouse's new design, Westinghouse moved to his demonstration train to drive home the point. Still, he had time since the trial report was still unofficial and awaited the formal authentication from the Master Car Builder's Association. Westinghouse took his train for demonstrations in the key railroad centers like Chicago, St. Louis, Albany, Cincinnati, Boston, New York, Philadelphia, Cleveland, and Pittsburgh. These demonstrations were attended by the public and engineering press, who delighted in the results. In particular, the executives of the biggest Eastern lines, such as New York Central and the Pennsylvania Railroad started to enter orders for the quick action triple valve. These railroad friends in the east, therefore, helped pressure the Master Car Builder's Association to review the Burlington trials in light of Westinghouse's new quick action triple valve. The committee filed the following:

> Mr. Westinghouse's wonderful optimism and confidence in the principle of air power, after the early failures were very marked. After the first collapse, in rapid succession, three different triples were invented. Finally came the gem, and then, greatest of all acts, it was exhibited all over the country in operation on a fifty-car train, making stops with no shock in the fiftieth car, not enough jar to upset a glass of water, a marvelous condition when one considers the wreck and disaster that used to take place in the fiftieth car.

This type of reaction and response to problems was characteristic of Westinghouse. Westinghouse was a problem solver, and he learned from experiments, as a scientist should. He never talked away from trials, but was enthusiastic about the ability to learn. Westinghouse saw trials and experiments as part of the design process, not verification alone. He could not be easily defeated because he was a professional engineer, and trial data was a means to improve design. The quick action triple valve gave Westinghouse a commercial victory, but it was a few months later that Westinghouse used what he learned from the Burlington trials to develop what many call his greatest invention — the friction draft gear.

The British and Burlington brake trials had generated much data on the science of braking. To most, the competition was the point of the trials, but as we have seen, Westinghouse used it for preventive design and failure analysis. This designing from device or system failures became a hallmark of Westinghouse. The trials showed Westinghouse a problem that was being overshadowed the focus on shorter stopping distances at higher speeds.

The physics of stopping a large train requires a transfer of energy. When a series of fifty cars are stopped, the cars are pushed together or crowded up in the stopping reaction. Trains had buffer springs between cars to absorb and buffer the force of this forward energy. The trials had shown this forward compression of fifty cars could be as much as sixteen feet over the length of the train. Once the cars were stopped by the brakes, the springs were compressed and the wheels locked. From a scientific view, the reactive energy is said to be stored in the springs. As the air pressure was released, freeing the wheels, the springs would recoil (snapping out) with a jerky motion that sometimes even broke the couplings. The train was often broken in two sections.

When Mr. Cassatt became President of the Pennsylvania, he noted that Westinghouse's greatest invention was the friction draft gear, yet it is one of his least known. It was clearly, of all his railroad inventions, the best example of applied science. Reading the patent, you see a maturing of Westinghouse the engineer. In the friction draft gear, Westinghouse reasoned from basic scientific principles and experimental observation, a new type of engineering device. In its development Westinghouse was drawn into several years of studying basic Newtonian physics. In the spring system, the energy of stopping is stored in the reactive compression of the springs, which is then released. The friction gear is part of the coupling and dissipates the energy of stopping through the dissipation of heat (via friction). Westinghouse's own description in his patent for the friction draft-gear shows his study of Newtonian physics:

My present invention relates to certain improvements in buffing apparatus designed to be interposed between a stationary and movable body, or between two bodies approaching each other either from opposite directions or between two bodies moving in the same direction, but at different rates of speed; and the invention has for its object a construction of buffing apparatus, whether applied to draw bars or buffers of cars, or for other purposes, wherein a frictional resistance is employed, either in combination with a spring resistance or alone, for the purpose of modifying the momentum and impact of the meeting or separating bodies.[29]

This invention and his approach show Westinghouse establishing the field of engineering, as we know it today. It marks him moving from being a great mechanic and inventor to an engineer. Westinghouse's drawing of the friction draft-gear is a piece of engineering art. These drafting skills go back to his father's workshop, but continued to improve over the years. For Westinghouse these drawings were art, and they are part of the legacy he left. Many geniuses like Edison, Bessemer, and Nasmyth left notebooks and hand drawings, but Westinghouse preferred the drafting board. It was in the drafting process that Westinghouse envisioned fit and interlocking of devices. Edison, for example, had an army of draftsmen to support the building of

29 U.S. Patent No. 391,991, October 30, 1888

his devices. Westinghouse differed from industrial scientists such as Edison, in that he used science to frame and reduce the necessary experimentation. Nikola Tesla noted the difference in approaches If Edison had a needle to find in a haystack, he would proceed at once with the diligence of the bee to examine straw after straw until he found the object of search. I was sorry to witness of such doings, knowing that a little theory and calculation would have saved him ninety percent of his labor." Westinghouse's knowledge and use of science saved him months over Edison's approach. Westinghouse was rare in his ability to dream, design, and build his devices. His brother Herman, a trained engineer, always admired George's engineering skills. His development of the friction draft-gear had brought him into the difficult field of thermodynamics. He attacked that science with passion, and later in life he often communicated with the famous Lord Kelvin on thermodynamics. Westinghouse actually progressed enough in thermodynamics to propose air conditioning and heat pumps.

The application of the friction draft-gear was slow in coming because the average train was still very short (less than twenty cars). The friction draft-gear would be critical to Westinghouse in expanding his air brakes through the freight market, which by 1900 would account for the 75% of railroad income. The Burlington trials were made with fifty car trains. In addition, cars were made of wood and much lighter than the steel cars that would become popular at the turn of the century. Still, it made inroads on the larger western freight trains of the time. For these long heavy freight trains, "it was impossible to make an emergency stop without parting the train by the breaking of coupler mechanisms." If this parting occurred on an uphill grade, part of the train would roll free. In 1888, Westinghouse was issued the patent, but a lot of testing was still needed to convince railroad men. Westinghouse hired a distinguished railroad man R. Soule to take the friction draft-gear into commercial development. Soule's work led to the worldwide commercialization of the Westinghouse friction draft-gear within two years. The combination of the draft-gear and triple action valve proved to be a system capable of stopping any length of freight train. The Master Car Builders Association made the Westinghouse Air Brake system the national standard in 1889. Soule as project manager finished the system design, allowing Westinghouse to move on to other interests.

This type of project manager hiring had already begun in other areas. Westinghouse realized that if he were to continue to invent in many diverse areas, he would need to turn inventions over to other managers for full commercial development. This practice is what would make Westinghouse companies so creative and innovative. Edison, for example, could never fully let go of his projects, and that limited him in the end. Westinghouse measured his success in terms of corporate success. The humility his mother had taught him worked well for his success. His engineers were always highlighted and retained ownership of their inventions. It was enough to advance his great companies. Soule, as project manager, started experimental trials with thirty car trains and was extremely successful. Still, railroaders were reluctant to spend dollars. The gear also was affected by rust, which required more development work by Soule. Westinghouse did work out an experimental deal with his neighbor Henry Clay Frick. Frick and Westinghouse set up a forty-five-car train to experiment with. The train ran coke of the Frick Coke Company from Connellsville, Pennsylvania to the Carnegie steel plant at Homestead. Soule worked endlessly to remove defects from the Westinghouse invention. By 1900 Bessemer & Lake Erie Railroad purchased it for 1000 steel cars, which represented the first big purchase. By 1910, the friction draft-gear saw a boom in application. By 1920, 90% of railroad cars had Westinghouse's friction draft-gear on them. Westinghouse had correctly foreseen the importance of the friction draft-gear as trains grew. Southern Pacific Railroad trials in 1908 proved that the successful operation of the Westinghouse air brake on long trains with heavy steel cars would require the friction draft-gear. With the friction draft-gear and the air brake, Westinghouse had changed the railroad industry. When Westinghouse started making air brakes in 1869, freight trains were 15 cars long, carrying 300 tons, and averaging 30 miles per hour. At his death in 1914 they were 130 cars long, carrying 7,000 tons and averaging 60 miles per hour. Statistics show by 1916, Westinghouse had put his brakes on over 3 million cars and 80,000 locomotives in this country alone.[30]

The decade of the 80s also pulled Westinghouse into the emerging field of electrical engineering. The impetus for this was Union Switch and Signal, which was pursuing electrical controls. Westinghouse was gathering some of the best engineers in the world at his Union Switch and Signal Company. By the end of the decade Westinghouse had developed and improved the interlocking switching to its modern format. The system was an electropneumatic system, where electricity activated valves and compressed air did the work. This system used a battery circuit through the rails, when a train entered a designated block; the circuit was short-circuited, activating pneumatic signals known as semaphores. In 1890 Westinghouse added automatic interlocking switches using electric current to activate hydraulic pressure units to open and close track switches. This represented an automated block signaling system no longer dependent on visual sightings or telegraphed messages. The automated electrical system took human error out, which accounted for over 90% of the accidents. Westinghouse could now add cybernetics to the sciences advanced. The last patent issued listed co-inventor J. G. Schreuder who headed the project for years. Schreuder rose to Chief Engineer and then Vice-President of Union Switch and Signal. Another great railroad engineer on the project was John Pressley Coleman. The evolved electro-pneumatic final patent was issued in February of 1891. This would be the breakthrough product that Westinghouse had envisioned al-

30 George Fleming, *Pittsburgh, How to see it* (Pittsburgh: G. Johnston, 1916), 213

most fifteen years earlier. Sales to rail yards and junctions took off and Union Switch and Signal would dominate the market for many years.

An interest in electricity had been sparked earlier in Westinghouse's companies. He had lured one of Edison's top mechanical engineers, H. M. Byllesby, to one of his construction companies — Westinghouse, Church, Kerr, and Company. Westinghouse, Church, Kerr, and Company was formed to install Westinghouse Machine's steam engines in factories and power plants. Byllesby had been lured easily away from Edison's dominance, which was an example of organizational problems at Edison's autocratic Menlo Park. Byllesby was one of the designers of Edison's first electrical power plant, New York's Pearl Station. Westinghouse had been impressed with Byllesby when he had visited Edison's famous light up of Menlo Park. Byllesby had become a frequent dinner quest at Solitude, and he spent hours talking to Westinghouse on the future of electrical power. Byllesby had the same love of mechanical drawing and quickly became a key technical consultant to Westinghouse. H. Byllesby was initially made vice-president of Westinghouse, Church, Kerr, and Company, but more importantly, he functioned in Westinghouse's informal electrical group. Byllesby was also working with Westinghouse Machine Company and Westinghouse's brother Herman on power generation stations. Byllesby was active in the professional engineering societies and networks of the time. Byllesby, having been a Colonel in the Civil War, brought strong managerial skills to the Westinghouse organization. Byllesby illustrated one of Edison's biggest weaknesses; the domination of employees, not allowing them to develop as individuals. Employment at Westinghouse offered unlimited potential for professional growth, and that attracted the best minds of the era. This difference in human relations management would be a major factor in Westinghouse's victory in the war of the currents.

Westinghouse by the mid-1880s was one of Pittsburgh's richest men. When the Westinghouses moved into Solitude in 1871, they were house poor with few furnishings. Solitude became an important daily retreat for Westinghouse as the years went on. They added many rooms and improved the grounds. Mrs. Westinghouse loved to garden and maintained both flower and vegetable gardens, and the house included a large conservatory for indoor gardening. Furnishings were added as well as many European souvenirs. Still, having a place of honor was George's boyhood stuffed hunting trophies. Westinghouse applied all the latest technology as it became available. He was one of the first in Pittsburgh to have a telephone and he had direct line connections to his plants and offices all over the city. Mrs. Westinghouse struggled with many ailments and often had reduced mobility. She was fond of hiring laid off Westinghouse employees to keep the grounds maintained. One of their favorite additions to the house was a large high-ceiling dining room. Another favorite extension was a large porch for dining and entertaining. The Westinghouse nightly dinners were known throughout the region. Westinghouse preferred to invite his business associates to dinner with their wives instead of the all men's meeting at places like the Duquesne Club or University Club. The dinners had many famous quests, such as Henry Clay Frick, Prince Albert of Belgium, Lord Kelvin, great scientists, European princes, generals from all over the world, and several American Presidents. Often he would mix these great guests with the train crew that would be waiting late to take them home. The train station was a short walk, but Westinghouse built a tunnel from the house to the station, so he could come and go with little publicity. Westinghouse took the train to his plants and office every day. For longer trips he had his personal Pullman coupled and waiting at the end of the tunnel. He always avoided the press except in the rare instances of new product that he wanted to promote. He was a great neighbor, often having their homes wired by one of his maintenance crews. He loved to drop in on Henry Clay Frick for a weekend game of poker. Frick's poker games often had many distinguished quests, such as future Senators, Congressman, and secretaries of the Treasury. Mrs. Westinghouse and Mrs. Frick also shared their love of gardening, even though, the almost continual darkness of Pittsburgh made gardening difficult. Between Frick, Carnegie, and Westinghouse, five American Presidents had visited the neighborhood.

During the 1880s, Mrs. Westinghouse's health deteriorated. She had several unidentified illnesses on her European trips, but physicians attributed her condition to the smoke and dust of the Pittsburgh air. Such problems from the sulfurous and dust laden air of Pittsburgh were not uncommon. At their suggestion, the Westinghouses spent some summers in Berkshire Hills, Massachusetts. Berkshire Hills and its summer cottages had become the resort of the Gilded Age. Artists and writers, such as Melville, Longfellow, Holmes, Hawthorne, and Beecher, had cottages there in the 1850s and 60s. By the 1870s, Berkshire Hills had become the home of industrialist's mansions, such as Carnegie and Vanderbilt. Presidents such as Taft, McKinley, Cleveland, and Roosevelt used to summer vacation there so they could do a little fund raising as well. Mrs. Westinghouse fell in love with the area from these cottage vacations.

Eventually they bought property at the corner of the towns of Lenox, Lee, and Stockbridge in 1887. The following year they added more property to what they would call "Erskine Park" after Mrs. Westinghouse's maiden name. The original name was the Schenck Farm. The farmhouse was expanded into a beautiful mansion. The marshy acres were drained and an artificial lake (Laurel Lake) developed. Mrs. Westinghouse became extremely active in the extensive landscaping and building of a sixty-acre park. House additions included dining rooms, a gym, bowling alley and a billiard room similar to the one at Solitude. Visitors that noted the distinguishing feature of the grounds was the white carriage paths, the brilliant white being derived from local crushed marble. A lighted tennis court was added as well. Lighting and telephones were added in1889 when it was in its infancy. Electric lamps were everywhere reaching 1500 in number by 1910. Like at Solitude, Westinghouse never put wires behind the wood paneling or ceiling panels, preferring to have easy access to redesign the lighting. Telephones were added to all rooms. As noted they were in daily contact when George traveled, following an earlier habit of telegraphing daily prior to the invention of phones. George, continued to make Erskine Park an electrical and mechanical wonder. Eventually, it had its own power plant to produce electricity using Westinghouse steam engines to run dynamos. Westinghouse even designed a special lighting rod system for his new mansion. He had first designed an elaborate lighting rod system at Solitude, which had become a prominent feature in Homewood. From 1890 this would be the summer home.

Erskine Park seemed to cure Mrs. Westinghouse, and its development was her project. While many industrialists of the time had huge winter and summer mansions, the Westinghouses were different in that these were true living quarters. Westinghouse was a homebody and brought his business home whenever possible. The walls were adorned not with great art but beautiful portraits and pictures of family and places visited. Her gardens were extensive and a gardener was hired. She worked hard on the smallest detail of landscaping, and was honored often by the Lenox Horticultural Society. Her gardener Edward Norman also won many awards for his landscaping of Erskine Park. The cleanliness and glow of this estate was in stark contrast to Solitude in Pittsburgh. Mrs. Westinghouse loved to wear all white dresses in place of her black Pittsburgh wardrobe. She also purchased white mahogany furniture, something that would never been done because of the constant black dust of Pittsburgh. Marguerite also made more extensive use of silverware, which the sulfuric air of Pittsburgh darkened daily requiring constant polishing. The many acres had been a working farm. Mrs. Westinghouse converted some of the buildings into additional recreational areas and living quarters. She also started several new hobbies such as raising horses. At Solitude, she had purchased some of the best horses for her carriages in the city but lacked the space for raising them. She and George built a working electrical dairy as her herd of cows went over thirty. She was fond of entering her horses and cows in local contests. Erskine Park became a real tonic for the sickly Marguerite, and she seemed almost reborn. George as always got involved with the neighbors. Every Fourth of July, he personally put on a fireworks show. After the Panic of 1907, he retreated to Erskine Park to pursue a number of technical hobbies. On Laurel Lake he built what is believed to be the first electric-powered boat, which he re-charged in his new boathouse. He designed a pumping system on the lake to supply water to the towns of Lennox and Lee water in times of drought. Erskine Park by 1914 had become his "Monticello" both physically and mentally. Like Thomas Jefferson, he loved to invent and build devices to aid living there.

Erskine also catered to his interest in bowling and golf. In golf he maintained membership at Pittsburgh's Oakmont Country Club and New York's Sleepy Hollow Country Club. Golf at the time had become very fashionable with America's industrialists; men like Carnegie took the game up in his 60s.

Westinghouse had always enjoyed his neighbor Henry Clay Frick's bowling alley in Homewood, and added one to his Erskine mansion. He added a golf course, which turned out to be a great attraction for his quests. The clubhouse and lounge became a place of business after a round of golf. Westinghouse added phones so that his quests could get in touch with their offices. Later in life, Westinghouse enjoyed fishing in the afternoon at Erskine. While George's father often questioned his son's lavish homes, Westinghouse homes were just that. He and Marguerite decorated, not with the expensive art of the Gilded Age millionaires, such as Morgan, Frick, Carnegie, and Vanderbilt, but with family pictures of their travels and small souvenirs, somewhat surprisingly since Westinghouse rarely allowed the press to photograph him or his family.

Chapter 7. A New Interest

> The Electric development we know today would have long halted without his daring and resourcefulness
>
> Lord Kelvin

Westinghouse is best known in relationship to his conquest of alternating current. This interest in electricity evolved as he tried to apply electricity to railroad brakes and signaling, but it was a natural for Westinghouse who was extremely talented at transforming abstract scientific ideas into practical engineering. Electricity also suited Westinghouse's need for a cybernetic system, which used feedback as part of the control system. Westinghouse's first interest in electricity had come from his involvement in the 1876 Philadelphia Centennial Exposition. The exposition was known for its display of giant steam engines. In the side exhibits, several electrical generators known as dynamos were on display. The dynamo generated an electrical current by moving a magnet back and forth in a copper wire coil. Dynamos were driven many times by Westinghouse steam engines, which represented a large market for Westinghouse Machine Company. These dynamos used this type of direct current (DC) to operate arc lights. These bluish lights were considered curiosities. These lights came from the arcing of electricity between two electrodes. Most electrical applications of the time were direct current (DC) from batteries. The streets outside the Exposition were illuminated by gaslights. Civil War veterans known as "Lame lamplighters" did the lighting of the lamps often. Edison had not yet started his experiments on incandescent lighting using direct current. George Westinghouse even then was fascinated by alternating current as a potential application for steam engines. Providence seemed to be pushing Westinghouse in the pursuit of alternating current from 1876. Westinghouse's interests by the late 1870s were converting on electrical technology. He was pioneering arc lighting in his factories, using electricity to activate railroad switches, and adding electrical controls to his air brake systems.

The 18th century is a movie of the evolution of electrical power. It opens on the beginning days of 1800 with Alessandro Volta reporting to the Royal Academy on his battery, and ends with a huge Westinghouse turbo alternator supplying alternating current to New England. DC current of the simple Galvanic cell or battery type power was the power source for the first 90 years of the century. The big breakthrough in electrical technology was the dynamo of Michael Faraday in 1831. Faraday's simple discovery was that an electrical current is generated if a conductor is moved back and forth between the poles of a magnet. Faraday's early experiments were studied by Westinghouse as a starting point for his entry into the field. Direct current from a battery moves in one direction only while alternating current reverses direction. Hippolyte Pixii invented a device known as a commutator, which took Faraday's moving conductor current and changed to direct current. This allowed a dynamo to generate steady, usable direct current. One of the major differences between AC (alternating) and DC (direct) current is how you can deliver it. High voltage (pressure) AC current could be distributed over very long distances. The characteristic that measures the pressure of the current is voltage. DC current is by its nature low voltage and lacks the push to deliver it long distances. One mile was about the practical limit for DC current. AC current is by nature produced at a high voltage that can push it many miles. Practically, a transformer must step down high voltage AC current to a lower voltage for use. This simple engineering principle is analogous to the gas delivery system designed by Westinghouse, in which high-pressure wellhead gas is stepped down in pressure by increasing the delivery pipe diameter. It was in this very principle that the secret to Westinghouse's success laid. Westinghouse's use of AC current earned it the name "Westinghouse current." AC offered major cost advantages, not only requiring many fewer power transmission stations, but a third less copper wire. It was only Edison's stubborn belief in DC that slowed AC use.

To start the story, it is necessary to go back to Westinghouse's inventive rival — Thomas Edison (1847-1931). As a boy, Edison had a passion for chemistry and electricity, which may have predestined him to champion DC current. His dangerous childhood experiments generated electricity from chemical cells consisting of zinc, copper, and sulfuric acid. He copied the work of Volta in producing these "wet" batteries. The real danger of these experiments was not in the low voltage current produced, but in the sulfuric acid used in the battery. Edison had a love for chemicals and their magic that remains unmatched. In the late 1920s as Henry Ford rebuilt the extensive laboratories of Edison in Greenfield Village, the last thing Edison gave to the exhibition was thousands of bottles of his cherished chemicals. He would spend most of his life trying to improve on the chemical battery. In fact, one of his last great inventions was the acid-less or as we call it today the alkaline storage battery. To supply his passion for chemicals, he made his money as a young boy as a telegraph operator.

As a telegraph operator Edison learned the usages of DC current to send messages over wires. Edison loved chemistry and experimented throughout his boyhood with electrochemical batteries to generate electricity for his devices. As a young boy he had read the two volume electrical classic by Michael Faraday, *Experimental Researches in Electricity*. George Westinghouse would start his study years later with Faraday's 1826 classic as well. Edison, however, was but a boy with no technical background when he started this monstrous text. This fascination with electricity and its uses lead to Edison's first invention that of the stock ticker. In 1876, the year of the Philadelphia Exposition, Edison opened his scientific factory known as Menlo Park. The persistent bulldog of an industrial scientist, Edison, set a goal of a minor invention every ten days and a major one each six months. In 1878 Thomas Edison started his campaign to develop the incandescent light in the same way Westinghouse approached such an endeavor with a search of everything written on the subject. He then visited William Wallace, who working with a DC current dynamo had developed an arc light. The arc light itself was less impressive to Edison then the efficiency of the DC current generator or dynamo, producing a higher voltage DC current than batteries. By 1879 Edison had a very efficient DC generator.

In the meantime, a University of Michigan electrical engineer, Charles Brush, commercially had developed a DC dynamo and arc light system. Brush as a boy had also read Faraday's text, while building telescopes and microscopes. Brush was different than Edison and Westinghouse in his formal training in electrical engineering. Brush being from the University of Michigan would start a rivalry with the Ohio State University in electrical engineering that would predate the great football rivalry of the two universities. Brush had pioneered arc lighting in 1876 prior to Edison's entrance into the field. On April 29, 1879, Brush used his DC dynamos and arc lights to light a park in Cleveland, Ohio. It was a modest setup of twelve lamps on wooden posts. Arc lighting was at its best in outdoor application. The dynamos were being driven by Westinghouse steam engines. The arc produced a harsh welder type light that would be unacceptable for rooms or reading. Arc lighting systems required almost daily replacement of the carbon arc electrodes by a "lamp trimmer." Arc lighting, however, got off to a fast start with San Francisco and Niagara Falls installing Brush systems by the end of 1879. Besides the harshness of the lighting, another problem was that the carbon electrodes were consumed by the electrical current requiring constant replacements. Edison seemed to have the better path of incandescent lighting.

Other early competitors in the lighting field were Elihu Thomson and E. J. Houston. Thomson was more like Westinghouse having been interested in mechanical devices as a boy. He made many devices as a boy including a camera and generator. His love of chemistry, however, defined his career as a Professor at Boys' Central High School in Philadelphia. Thomson began building DC current dynamos with fellow professor E. Houston. Thomson would use his dynamo and nine arc lights to light a Philadelphia bakery in 1879. Thomson never proved to be much of a businessman and Charles Coffin, a successful shoe salesman, took over his operation of Thomson-Houston in 1883. Thomson stayed with the company as the technical vice-president. Coffin initiated a policy to establish a monopoly in the arc lighting industry. By 1890 Thomson-Houston controlled over two-thirds of the arc lighting business. Thomson-Houston also moved into AC and DC power generation, but it would later be absorbed into General Electric.

These arc lighting successes over shadowed the 1878 formation of Edison Electric Light Company in late 1878. The partners of Edison were Norvin Green and Tracy Edison of the telegraph industry. The third partner was Egisto P. Fabbri, a partner in J. P. Morgan's company. The capital would launch Edison on a successful campaign to develop a filament to perfect the incandescent bulb. More ominous was the involvement of J. P. Morgan, who ultimately would undo Edison and Westinghouse. Edison famous quest of endless trial and error experiments to find a filament material took almost two years. Edison had gone against the trend with his research in 1877, and it had paid off with a practical incandescent light for homes. In October of 1879, he found success with a carbonized bamboo filament. A month later in 1880, a special train of the Pennsylvania Railroad brought hundreds of reporters to see a lighted Menlo Park. Later in January, Westinghouse hooked up his Pullman car to visit Menlo Park and observe Edison's light demonstration. Edison treated Westinghouse with great courtesy. Westinghouse, not yet fully involved in the struggle would feel this as the first blow of a long battle, as his gas company stock dropped in anticipation of incandescent lighting replacing gas. Edison Electric stock shot up from $10 to $100 as Edison's invention filled the headlines. One of Edison's unheralded strength was his love and manipulation of the press. By February, as sober scientists evaluated Edison's system, the stock returned to $20. The first commercial use of the Edison system was on the steamship S. S. *Columbia* in1880. Edison continued to win over the public with his circus like demonstrations, but commercial lighting applications were coming to arc lights.

It was a slow but steady path for Edison from 1880 to 1886. Edison's incandescent light could supply more light at a lower price than gas for the homeowner. German electrical engineer, Werner Siemens refused to purchase Edison's incandescent light licenses in 1881, saying arc lighting was the future. Westinghouse as well in 1881 felt arc lighting was the future. Edison would of course be proved right on the incandescent light, but it would take him also twenty years to acknowledge he was wrong about powering them with DC current. Arc lighting was simple and easy to install and maintain. Incandescent lighting had one other very important supporter in J. P. Morgan. Edison's reputation and Morgan's money would turn the tide to incandescent lighting.

Historians had ignored Westinghouse's early experiments and role in the commercialization of direct current arc lighting. Westinghouse had been using Brush arc lighting at least in his machine plant in 1878. As early as 1879, Westinghouse was experimenting with lighting systems.[31] His main interest at the time was to light his factories at Liberty and Twenty Five Streets, but his brother and other business had caught Westinghouse's enthusiasm for this new technology. Herman Westinghouse, John Caldwell, and Ralph Baggaley approached him on forming a company to sell lighting units in 1879. At the time Westinghouse was lighting his factory with forty light Brush arc machines powered by Westinghouse Machine steam engines. He had even run wires to light the near by Pennsylvania Railroad yard. Herman Westinghouse saw it to be a great application for a new high-speed steam engine he was working on. The men came together to form Allegheny County Light Company with H.H. Westinghouse as president, C. Magee as vice-president, John Caldwell as treasurer, and Ralph Baggaley as a director. Westinghouse purposely stayed in the background; he did not want to drive Edison Electric into buying up competing patents. Westinghouse was a major investor and consultant. One of their first customers was Joseph Horne's Department Store, which installed arc lights for evening shoppers. Each evening a floorwalker checked the lights and notified the "lamp trimmer" to add carbons. The lamp trimmer reported the adds to the light company, and that has how Horne's was charged for usage. Another early customer was the Monongahela wharf market area. Success brought competition with the formation of Diamond Street Electric Light in 1882, which George Westinghouse bought out in 1884, incorporating it into Allegheny County Light. Westinghouse by this time was convinced that AC current would be needed to extend power throughout the city. That struggle would also involve the new incandescent light.

The struggle turned financially and commercially, as J. P. Morgan became the first in New York to install Edison's incandescent lighting at his mansion. While it was a huge publicity success, the actual operation was full of problems:

A cellar was dug underneath the stable which stood on Thirty-sixth in the rear of the house (Morgan's house address was 219 Madison Avenue), and there the little steam engine and boiler for operating the generator were set up. A brick passage was built just below the surface of the yard, and through this the wires were carried. The gas fixtures in the house were wired, so that there was one electric light bulb substituted for a burner in each fixture. Of course there were frequent short circuits and many breakdowns on the

31 Greater Pittsburgh Chamber of Commerce, *Pittsburgh and Pittsburgh Spirit* (Pittsburgh: Robert Forsythe Co., 1928), 203

part of the generating plant. Even at the best, it is a source of a good deal of trouble to family and neighbors. The generator had to be run by an expert engineer who came on duty at 3:00 P.M. and got up steam, so that at any time after four o'clock on a winter afternoon the lights could be turned on. This man went off duty at 11:00 P.M. It was natural that the family should often forget to watch the clock, and while the visitors were still in the house, or possibly a game of cards was going on, the lights would die down and go out. If they wanted to give a party, a special arrangement had to be made to keep the engineer on duty after hours.[32]

It was said Mrs. Morgan wanted to throw "the whole damn thing in the street, but by then Morgan owned about a third of Edison's stock. He went on to feature his electric home in national magazines. Morgan convinced William Vanderbilt to light his house, but a short circuit caused a fire in the art gallery and the Edison system was out in the street. Morgan's stock interest grew and so did his personal interest in Edison's company. Edison with Morgan's help got the contract to illuminate Manhattan, but before it was even finished, a new challenge came from Europe. The low voltage, direct current he used, limited Edison's system. It could be profitable in densely populated areas where the cost of copper wire and generating stations would be spread out among many customers. In 1883, two engineers, Lucien Gaulard of France and John Dixon Gibbs, invented a transformer that would allow high voltage alternating current to be transmitted low distances and stepped down then for customer use. This new Gaulard-Gibbs transformer could send current several miles. Now Morgan was concerned, as his European associates warned that the future might go beyond to alternating current. Morgan tried unsuccessfully to interest Edison in looking at alternating current. Edison was adamant that his company had to use DC current.

The battle of the 1880s and early 1890s was not between DC and AC current but arc lighting and the incandescent lighting of Edison. Realize that the Edison system was a DC incandescent lighting system from 1878 to 1894. Really arc and incandescent were two separate markets by 1892. Both were growth markets. Arc lighting was just that the light was created by the electrical arc between two carbon electrodes much like arc welding today. The electrodes were consumed in the process. Arc lights were great for outdoor applications, such as streetlights, because of their simplicity, power, and brightness. Arc lighting also offered much more candlepower than gaslights, and could light railroad yards and parks. Arc lighting was an extremely profitable business, which had attracted Westinghouse to forming his Pittsburgh lighting company in 1880. Arc lighting would be a growth market into the 20th century. In 1881 there were 6,000 arc lights in service, by 1886 there were 140,000 in service, by 1895 there were 300,000 arcs in service, and by 1902 there were 386,000 in service. While technically, arc lights could be run by either DC or AC current, they tended to be DC. While Edison and Brush

32 Ronald Clark, *Edison: The Man Who Made The Future* (New York: G.P. Putnam's Sons, 1977)

arc lights were DC by 1888, Westinghouse had developed a fully functional AC arc lighting system (patented in 1890). Incandescent lights also could be powered by DC or AC. Edison's system was DC, and later Westinghouse would challenge it with AC because of its lower cost.

Incandescent lighting (while supplied by DC or AC power) was a softer light closer to a candle or gaslight. It lacked the strength or brightness for outside applications. The incandescent light offered better, more comfortable lighting for homes. Also incandescent lighting proved superior in factories where a softer light allowed for better defect inspection, color distinction, and texture. Early studies in factories showed a productivity increase with incandescent lighting. The market for incandescent grew at exponential rates as well. In 1882 there were 29,192 in service, by 1886 there were 181,463, and by 1888 there was 434,654 in service. Edison initially controlled the incandescent lighting market through his patents and superior product. There were competing incandescent lamps of Stanley and Weston. Edison had a major cost advantage over the competition and superior lamp life. Life studies done by the Franklin Institute in 1885 showed that in 1065 hours (approx. 6 months), the Edison lamp would fail at a rate of 1 out of 21, the Stanley at 17 out of 24, and Weston out of 22. Incandescent lighting also offered an advantage over arc and gas in industries where explosions were possible, such as flour and chemical industries. Incandescent lighting clearly would win the battle, but would the power supply be DC or AC.

A few minutes are needed on Morgan since he would ultimately impact the destinies of both Edison and Westinghouse. J. P. Morgan came from one of America's oldest and distinguished families. His father, J. S. Morgan, was a merchant banker whose primary focus was short-term trade finance. It was a cutthroat business that allowed merchants to sell their bonds or paper certificates to Morgan at a discount to raise cash. Another part of J. S. Morgan's was to supply credit to American companies in Europe actually as an underwriter. J. P. Morgan moved into his father's business in 1864. Morgan became the master of Big Business through control and interlocking boards of directors. These types of financial arrangements became known as "Trusts." In addition to these trust arrangements, Morgan still functioned as a merchant banker controlling credit. The power and control of Morgan was unbelievable. Even the government, prior to the Federal Reserve Act, depended on Morgan to keep the country fluid. He could also use "financial panics" to take over companies. In those days a bank could call in short term debt. If the company did not have the cash, the bank could end up owning the stock and bonds of the company. History now sees him as the de facto central banker of the United States for the Victorian period. His overriding belief was that competition was wasteful and destructive. He controlled industries to avoid competition and maximize corporate profits. Companies that got in his way could be subdued by drying up their credit in difficult times (known as Panics). Historian Charles Morris described this best:

Often strangely inarticulate, as if rendered speechless by the titanic fulmination in his breast, he railed against the madness for progress and change that wiped out perfectly respectable business of perfectly decent gentlemen, against the gale winds of technology that turned economic assumptions upside down and made it impossible for his clients to *pay their bills!* Over the course of forty years, he eventually succeeded, as no one else in imposing his own iron will on the American economy, reining in the competitive free-for-all, and setting rules and boundaries that held sway for a half-century after he died.[33]

No single American businessman ever had the power of Morgan over the economy. His strength was undeniable. At the time of Westinghouse Electric's fall in 1907 may well have controlled 40% of all capital related to industrial, commercial, and financial industries. Morgan's 1901 formation of United States Steel was the largest capitalization until the late 1980s. Morgan's trust arrangements of banking and business were unequaled until the zaibatsu of prewar Japan and its restructured postwar keiretsu. Morgan often rationalized that his approach of capitalism maximized shareholder profit. Morgan power for years went unchecked; refusal to place a Morgan man on the corporate board usually resulted in capital starvation and ultimately receivership. Westinghouse held Morgan off longer than anyone (from 1880 to 1907). Ultimately, it took two major panics and an army of allied bankers to do the job.

Probably only one American ever backed down Teddy Roosevelt, and that man was Morgan. In fact, no American president could afford to stand up to Morgan. Morgan was the government's financier. At least Roosevelt saved face; Morgan had pushed President Cleveland across the board in the gold panic of 1893. Morgan's extensive ties with European bankers such as the Rothschilds made him the US government's main source of credit. Morgan maintained the role of the American King. He bullied his way into President Cleveland's office in 1895, and made John D. Rockefeller wait outside his for hours unacknowledged. He controlled the railroad, steel, oil, electrical, and banking industries, as well as the world gold market. Many would add the US Government. Morgan would tame America's great industrial lions — Carnegie, Vanderbilt, Rockefeller, Schwab, and Westinghouse. Only Westinghouse refused to kiss the ring by not cooperating in his arrangements.

Edison on the other hand was a willing victim. Morgan had helped finance Edison initially, and Morgan lined up the New York customers for his Pearl Station generating plant. On September 4, 1882, Thomas Edison watched his system light lower New York from J. P. Morgan's office. It was the completion of a goal Edison had written in his notebook in 1877. By the end of 1882, the Pearl Station was lighting 3,400 lamps for 231 customers. By the next year it was lighting 14,000 lamps. Edison was also constructing a central power station in Detroit (Henry Ford would work as an engineer

33 Charles Morris, *Tycoons* (New York: Times Books, 2005), 28

there). Edison worked throughout 1882 to 1884 to address the geographic limitations of his direct current system. The direct current system had a drop in voltage as the distance expanded. To correct this required thick copper wires at the generator tapering off to the end of the line. The cost of copper inhibited the Edison direct current system. He designed a central feeder station concept to overcome the voltage drop and copper usage, while much lower cost it would be still more costly than alternating current systems. After that 20 mile distribution became possible with central and feeder stations. Profitability rose, as did Morgan's share in the company. Electricity found new applications such as in streetcars, subways, and electric trains. By 1887, there were 121 Edison power-generating plants in America and Europe. Edison basically sold most of his control of Edison General Electric in 1888 to finance other projects.

In 1888 J. P. Morgan was the largest stockholder of Edison Electric. Edison Electric remained the largest and most integrated producer of power and electrical lighting systems. Edison controlled enough stock and had friendly stockholders to block Morgan from imposing management or forcing Edison to consider AC current. Morgan, however, had started to plan reorganization at his library without Edison's knowledge. Thomson-Houston was a major competitor causing price problems for Edison Electric in the marketplace. Brush Electric in Cleveland represented a third competitor in the lighting area. Westinghouse distributed Brush lighting systems in Pittsburgh. Westinghouse Machine Company was a major producer of the steam engines needed for electrical generation for all the lighting companies. Westinghouse sat in Pittsburgh flush with cash from his many companies, and a declared mission to move into the electrical business using the competing AC current system. In addition, there was growing competition in lighting from United States Lighting Company and Consolidated Electric. All of this competition in lighting came from competing incandescent lamps, such as the Sawyer-Man, Swan, Stanley, and Weston. The whole industry was mired in lawsuits and patent battles. In fairness, I'm over simplifying the competitive nightmare that existed. If there ever was justification for a trust, the electrical industry of 1888 was it! It was legal driven innovation gone crazy.

For Morgan, this was capitalism at its worst locked in inefficiencies, legal battles, and rising costs. Morgan's natural tendency was to replace this with a trust controlled by the investors, an electrical trust merging the major companies or some form of banker-controlled directorships. Thomas Edison already suspected he was losing too much control, and was not open to a merger that would dilute his control further. Just prior to the reorganization of Edison Electric, Morgan sent agents to work on all the corporate players to consider the formation of a trust. If Morgan could have brought Thomson-Houston and Westinghouse into the fold, the others could be crushed, but he also had to deal with Edison internally. Morgan tied to play Westinghouse against Edison. His overture to Westinghouse came through a lawyer that Westinghouse had used in patent reviews by the name of Doctor Otto Moses. The Associated Press ran articles of unknown origin about a possible trust. Westinghouse suspected that Edison was also being manipulated, and wrote a letter directly to Thomas Edison on June 7, 1888:

Dr. Moses came to Pittsburgh some weeks ago with reference to a scheme for the consolidation of all electric light companies in some form of trust. I refused to have any thing to do with it, and told him that I saw no reason at all for our combining with a lot of people who had nothing to give.[34]

Westinghouse also rebuffed any trust talk in a letter to the editor of the newspaper that carried the story. It was Morgan's first attempt to win over the ethical giant. It would launch a twenty-year effort by Morgan to bring Westinghouse Electric into the trust.

The failed initial takeover of Westinghouse, forced Morgan to deal with Edison in his own company. In 1890 Morgan put his own man in charge of Edison General Electric, Henry Villard. Henry Villard, a Morgan man, had been President of the Oregon Railway that gave Edison his first commercial order. Villard, as President of Northern Pacific Railroad, had also underwritten three miles of electric train track at Menlo Park. When Villard took over the internal struggle for alternating current began. Morgan's associates were now very concerned about the possible challenge of alternating current to Edison's DC current system. This was about economical distribution since the incandescent lamps would work on either DC or AC current. Edison continued to fight its use and tired of hearing about the developments going on at Westinghouse Electric, but he was coming to realize that he was not in charge. For Edison it was becoming personal between him and Westinghouse. Villard had brought the leading expert of alternating current to Menlo Park. His name was Nikola Tesla, but Edison had forced him out. Morgan and his associates were determined to head off any competition between companies and AC/DC currents. Finally, in late 1891 Villard requested in writing that Edison visit Westinghouse. Edison lodged a terse reply:

I'm very well aware of his resources and plant, and his methods of doing business lately are such that the man has gone crazy over the sudden accession of wealth or something unknown to me, and is flying a kite that will land in the mud sooner or later.[35]

> Morgan sent the final message to Edison that he was no longer in charge a year later when the company name became General Electric through a merger with Thomson-Houston Company. Edison and the company continued to battle DC current publicly and Morgan attempted to bring Westinghouse under control by trust arrangements. Uncharacteristically, the two dueling inventors were holding up Morgan in implementing his overall control of the industry.

Morgan loved new technology, but he, like many even today didn't understand electricity. Personally, he didn't care much for Edison, but Edison had

34 Letter George Westinghouse to Thomas Edison, June 7, 1888, Edison Paper Archives, Rutgers University.
35 Clark, 160

a track record and star status. He couldn't understand Westinghouse, but he respected his engineering skills. In 1880 Edison Electric's own numbers suggested that to light 8,000 lamp circuit covering nine city blocks would require eight hundred thousand pounds of copper at a cost of $200,000. For two years Edison redesigned and invented a central-feeder station system that had brought that cost down to $30,000. It was an impressive piece of engineering and helped ease concerns at the House of Morgan. Furthermore, Morgan had debt control of Edison's company, and eventually, he would force his will. Morgan started to look at moving into the copper market as a defensive strategy. Morgan's interest in the technology did keep Edison Electric people, particularly in Europe watching AC developments, and it was in Europe that Westinghouse initially entered into competition with Edison Electric. Westinghouse had a solid DC current distribution system for Pittsburgh, but he dreamed of a wider distribution network, and that would require AC current technology.

The Gibbs-Gaulard transformer had allowed high voltage AC current to be stepped down to the usable voltage range after long transmission. This was the key which would unlock AC power. AC high voltage gave it a tremendous advantage in delivery, but when it got to the lamp that high voltage would burn out the filament — it needed to be stepped down to a lower voltage. The Gibbs-Gaulard transformer was a crude invention with lots of problems, but Westinghouse's people had quickly envisioned its future. Edison ignored its future by focusing on its initial limitations. In 1885, a new buzz was sweeping Europe about an improved transformer. Three Hungarians, Karl Zipernowsky, Otto Blathy, and Max Deri produced an improved transformer (Z.B.D. system) that could transmit a half-mile further than Edison's DC current. It was heralded at the Bucharest Exposition in 1885. The word quickly got back to Edison and Morgan. Agents were dispatched to Europe and an option to buy the patent was purchased by Edison Electric for $5000, Edison and his supporters would hold up the deal. The new AC transformer fired the imagination of European engineers and inventors, even with the safety warnings of Werner von Siemens and Lord Kelvin.

It may seem strange that such a great inventor and "wizard" would have a technical block to alternating current. The fact is that AC and DC current actually represent the functions of the two sides of our brain. DC current can be visualized readily, as water flowing through a pipe. Right side geniuses can do amazing things when they tap into their capacity to visualize, and that was Edison. Using his visualization of DC current, he applied it to endless inventions. Alternating current offers a different problem, it cannot be readily visualized, and it is theoretical by its very nature. This left-sided abstraction can be a difficulty for even a genius like Edison. Westinghouse, the true engineer, could mentally visualize such abstractions. Edison's biographer, Robert Conot, put it best:

> His [Edison] entire experience had been with direct current. He could visualize the action of a direct current, but the action of an alternating cur-

> rent was something that he could not comprehend — water could not be made to go back and forth in a pipe; and, even if it could, it would be useless for performing work. Because of the orientation of his mind, he had allowed the development of the transformer to slip by him while he struggled with less radical solutions to the problem of transmission.[36]

Had Edison fully comprehended AC current theory, he would have started the electronic era decades earlier. Edison correctly observed and described what has become known as the "Edison Effect." Edison actually called it an "etheric force." Unknowingly Edison had discovered electromagnetic radio waves, but failed to grasp electronic theory. Had Edison perceived what he so accurately described, he would have pioneered radio, television, X-rays, and vacuum tubes for electronics. Edison, however, was bothered by the idea of electricity as an electromagnetic wave; he preferred to see it as a force. Electromagnetic wave theory required an abstract view of electricity versus the pragmatic view of a master electrician. This was the same reason why Edison loved his hand drawn three-dimensional drawings and hired draftsmen to convert this to two-dimensional engineering drawings ("blueprints"), while Westinghouse loved to go directly to the engineering drawings. Westinghouse mentally could deal with technical abstractions and bring them to useful applications. Westinghouse could visualize the application of alternating current like Edison could DC current. Westinghouse did not have difficulty in seeing electricity as a wave versus a force, allowing him to grasp difficult electrical theory.

As noted Westinghouse had causally came into the "war of currents." His first encounter with lighting came through the formation of Allegheny Lighting in 1880 and later that year with dynamo engine sales to Brush Electric from Westinghouse Machine. Westinghouse Machine was formed to produce the high-speed steam engine of his brother Herman Westinghouse. Westinghouse Machine's first contract was to produce steam engines to drive Brush Company dynamos. This was in 1880; during the decade Westinghouse would become the dominant steam engine manufacturer for electric generation including Edison DC generation. Westinghouse foresaw the business potential of electrical generation long before he was concerned about the advantages of AC current. During the early years of Westinghouse Machine, his brother Herman had developed a friendship with a number of electrical engineers and inventors. William Stanley was one of these, who had invented a self-regulating dynamo for Brush Electric. Another was Elihu Thomson who had done work on direct current dynamos and incandescent lighting before Edison. Direct current dominated those early years because electrical devices of the time required direct current. Herman continued to experiment and work with Brush Electric as George was tied up with the gas company and Union Switch and Signal. Herman had even talked Stanley into working with him. Herman Westinghouse with George founded the New York engineering firm of Westinghouse, Church, Kerr and Company.

36 Conot, 253

Eventually, Westinghouse was able to convince Stanley to join the Westinghouse research team at Garrison Alley in1884. Stanley was a restless eccentric, who liked to be a loner. Fellow workers of the time described him of having a "nervous disposition." Stanley also seemed to have a much higher opinion of himself then his fellow workers. George Westinghouse demonstrated great tolerance with eccentric inventors like Stanley. Fellow workers suggested he spent more time in Berkshire Hills than Pittsburgh. He didn't fit in well with any team effort, but he had been a visionary of AC current. Herman Westinghouse saw Stanley as a utility man that could contribute to Westinghouse Machine's engine business as well as George's electrical R&D work at Switch and Signal's Garrison Alley. Stanley got off the train at Pittsburgh complaining about the smoke and dirt. At bit of a health nut, Stanley loved to summer with the rich in Berkshire Hills.

George Westinghouse entered the field more through his Union Switch and Signal Company versus Westinghouse Machine. Stanley worked with Union Switch and Signal as well, since the Westinghouse corporate lines were not fully drawn yet. Switch and Signal was working on applying electrical devices to this growing field, mainly through Westinghouse's experimental Garrison Alley lab, then known as the electrical department of Switch and Signal. Garrison Alley was as close to Edison's Menlo Park as possible in the early days. It worked on numerous projects of interest to George Westinghouse. Union Switch and Signal already had European electrical engineers, Guido Pantaleoni and Albert Schmid, in its employ. Electric lights could offer a significant advance in the art of railroad signaling. Pantaleoni was a bit of an eccentric himself, and had problems with Stanley. Both men had huge egos, but Westinghouse again showed his ability to house an unusual collection of personalities at Garrison Alley. Pantaleoni commuted from St. Louis, so like Stanley who commuted from New England, his disruptive influence was limited. Albert Schmid on the other hand demonstrated supervisory and managerial skills, leading to his role of Superintendent of the lab. By 1884 George was also exploring the technical literature on electrical devices and generation to aid in the advance of signaling. Westinghouse favored AC because railroad signaling would cover miles, but AC research was lagging in America. AC current was progressing in Europe on several fronts. Werner von Siemens had a prototype AC generator. He had read about the Gaulard-Gibbs AC transformer, which could step down high voltage AC current. Herman also was excited by the potential of AC current production and use, but a lot of Westinghouse's organization opposed AC. Westinghouse's patent lawyer and electrical engineer, Franklin Pope, opposed the move into AC experimentation, fearing it was too dangerous. Ironically, ten years later Pope died in a high voltage experiment.

In 1885, while Westinghouse engineer Guido Pantaleoni was in Europe on a family visit, he explored the Gaulard-Gibbs technology. Westinghouse's interest was inspired by the telegrams and reports from Pantaleoni. Westinghouse asked for both the Gaulard-Gibbs and Siemens devices to be sent to Pittsburgh. The Siemens generator could be connected to a Westinghouse steam engine, then the current voltage could be stepped up by the Gaulard-Gibbs transformer and then stepped down at the point of use. The Gaulard-Gibbs transformer arrived with one of their engineers, Reginald Belfield, who Westinghouse would hire. The Gaulard-Gibbs was to crude for application, and Edison Electric held the patent right option on the Z.B.D. transformer forcing Westinghouse into internal development. In 1885 Westinghouse assigned Stanley and Belfield to redesign these transformers for commercial use. Westinghouse now had a major research and development effort going on at Garrison Alley. He personally invested in this corporate joint venture $150,000. More engineers such as Albert Schmid and Oliver B. Shallenberger were committed to the AC current project. Shallenberger had been working on arc lighting applications for Westinghouse. Shallenberger was a gifted practical electrician with a temper, so he added to this temperamental group of innovators.

The receipt of two damaged Gaulard-Gibbs transformer turned out to be fortuitous. Westinghouse joined in with Schmid, Belfield, Pantaleoni, Shallenberger, and Stanley. Westinghouse was like the teenagers of the 1950s tearing apart a car engine. He loved this type of mechanical dissection. Once he had taken an employee a new type of Swiss watch and disassembled it to learn the secrets of its operation. Similarly, he once took apart an American made watch of his partner — Belfield. He loved to study these mechanical movements. For November and December of 1885, Westinghouse headed for Garrison Alley every morning to work on the tear down and analysis of the transformers. By the end of the day he would invite his tired engineers to dinner at Solitude and discussed progress to the small hours of the morning. Belfield would describe Westinghouse's participation as Mr. Westinghouse applied himself toward the production of a piece of apparatus which could be wound on the lathe, utterly discarding the unpractical soldered joints and stamped copper disks for the more commercial form of ordinary insulated copper wire. It took Mr. Westinghouse only a few days to design an apparatus which has been the standard ever since." By January they had designed a new type of AC transformer, and Westinghouse stood alone in its future application. Westinghouse, however, was fully convinced of the AC current. Eventually Stanley was persuaded to switch to AC, although, he had a fully designed DC system ready to be launched by the company.

In January of 1886, Westinghouse Electric was formed out of the experimental branch of Union Switch and Signal Company. When Union Switch and Signal moved to Swissvale, Garrison Alley became headquarters for Westinghouse Electric. The corporate charter included the statement, "to manufacture and promote the use of alternating current system equipment." The corporation was formed with a million dollars of capitalization. The directors and major investors were George Westinghouse, Herman Westinghouse, John Caldwell, John Dalzell, Frank Pope, John McGinley, C. Jackson, and Robert Pitcairn. Westinghouse retained more control of this company to assure his freedom to innovate and experiment. Pope represented a new partner. He was a patent lawyer and electrical expert from New York that Westinghouse had been consulting with. In addition, Pope had a national reputation and was editor of *Electrical World* for years. Herman Westinghouse was president of the company and Guido Pantaleoni became General Manager. Shortly thereafter H. Byllesby would become vice-president and general manager. William Stanley was named chief electrical engineer, and reporting to him was Oliver Shallenberger as chief electrician. The company took over the Garrison Alley plant and lab as Union Switch and Signal moved to Swissvale. Westinghouse started to research purchasing the patents for the Gaulard-Gibbs transformer to start out. Westinghouse had now formally entered the field. While Edison dismissed this as almost silly, Morgan watched with great interest, still trusting Edison's belief in DC. Edison was by 1886 solidly entrenched in both the American and European market. More importantly, AC power generation was blocked by the lack of an AC electric motor. The DC current on the other hand, was by 1886 driving streetcars and factories all over the world. Westinghouse split the research effort with AC power generation, AC incandescent lighting, and development of an AC motor.

Just as Westinghouse had launched what became known as the "war of the currents," he started a project at Westinghouse Machine Company that honed his mechanical and electrical insight. Westinghouse Machine had become one of the premier producers of steam engines for all purposes. President Herman Westinghouse's expertise in building these engines had brought Westinghouse Machine to the national forefront. The company launched an extensive study at customer operations on the efficiency of steam engines. Engineers with testing equipment were sent all over the world to study steam engines. No other company had even dreamed of applying science to the study of these behemoth engines as well as the smaller ones. The study published in *Practical Electricity* and *Scientific American* published the results in 1889.[37] While the results were withheld as to the companies tested in the study (because of the embarrassment), they were eye opening. One conclusion was "nearly all were wasting one-half of their engine power (or one-half of daily consumption of fuel) before commencing actual work, the product from which constituted the maintenance of the business." The study resulted in a training program for users similar to the one Westinghouse had designed for air brakes. Westinghouse engineers were available to customers to help them run their engines better. It represents one of the first examples of service engineering in industry. Henry Ford in his early days as a steam engineer would be one of the men who learned much from the service. Just as important was the fact that it changed the focus of Westinghouse to one on efficiency in both mechanical and electrical devices. That focus would later result in a competitive advantage in electric motors years later.

37 "Business Success," *Scientific American*, July 27, 1889, page 52

Westinghouse to a large degree would spend the balance of his life trying to improve engineering efficiency in a variety of devices. In the late 1880s, Westinghouse's vision of electrical appliances was beyond his competitors. He firmly believed that electric motors would replace steam driven power in all engine applications.

Westinghouse's many diverse interests started to come together with the emergence of Westinghouse Electric and Manufacturing Company. He had early on used electricity successfully in electro pneumatic braking devices and control. He obtained a patent for electro pneumatic control in 1881. Union Switch and Signal had brought him further into electrical control of signals, lights, and hydraulic valves by 1886. He had patented a telephone switching system in 1880, and a telegraphic relay switch in 1886. By 1887, Westinghouse was putting electrical arc lights into factories. All this before Thomas Edison had started his incandescent light experiments. The combination of much of Westinghouse's inventions could be seen in the cutting edge of the Pittsburgh fire department. In 1888, Pittsburgh's No. 1 Engine House was the pride of the city. It had an electrical alarm system with alarm boxes throughout the city. Prior to this, fire alarms were church and school bells. When an alarm box signal has activated, it went to a Central Alarm Office via telephone lines. The Central Office used another telegraphic relay to open inner doors of the horse stalls. With the opening of the doors, "86 horses, which, with the well-known intelligence of the fire horse, at once take their places. They are harnessed by catching the collar, which, with the rest of the harness, is suspended by cords running over pulleys to a counterbalance, and pulling it into position, when with two sharp snaps the harnessing is completed."

Chapter 8. The War of the Currents

> George Westinghouse was, in my opinion, the only man on this globe who could take my alternating current system under the circumstances then existing and win the battle against prejudice and money power. He was one of the world's true noblemen, of whom America may well be proud and to whom humanity owes an immense debt of gratitude
>
> Nikola Tesla

It may seem strange to most that Westinghouse wanted and even looked forward to what would become the "war of the currents." It wasn't about money; it was about achievement and winning. The war of the currents offered a technical challenge beyond anything he had faced in the railroad or gas business. For Westinghouse to even have such a great opportunity was viewed as a blessing. Personally, in 1887 money was pouring in from all his enterprises. Westinghouse Air Brake dominated the market, and Westinghouse Machine was shipping 45 big engines a month. The machine company had expanded into steam-powered hoists and overhead cranes. The National Railway Act of 1884 assured the air brake business for future decades. Westinghouse also had operations in 12 different countries. The challenge of electricity was drawing in Westinghouse when others would have retired. It was even more than a great technical battle; it would be with the world contender for its greatest inventor — Thomas Edison. This would not be personal for Westinghouse; it would always be the challenge and the quest for achievement that was the focus of Westinghouse. Westinghouse wanted a fair fight, but this would be his initiation into the world of Big Business. This would be different, it was a business he could not finance personally, and that would take him into the world he despised most — banking and finance. He would win the war but cross America's most powerful man in that victory, and J. P. Morgan would remember and wait.

It may help to summarize the situation of DC and AC current in the late 1880s. Edison's incandescent lighting and power generation system used DC current. DC current had serious transmission problems requiring excessive amounts of copper, and power stations every mile. For this reason Edison's DC systems were favored in densely populated areas, such as New York. DC held only one but important advantage that of a superior commercial electric motor. The DC motor favored DC current's use in streetcar systems, such as those of Thomson-Houston. Incandescent lighting could operate on either AC or DC with certain modifications. Edison held most of the patents relating to DC current, while Tesla owned those for AC. High voltage AC current presented some safety problems and the AC motors were crude. Finally, Thomas Edison opposed the use of AC systems for any of Edison Electric Company projects. Edison would hold the upper hand until 1893 and the World Fair.

From 1880 to 1884, J. P. Morgan poured money into the Edison Company with success and profitability, while Edison directed the company technical campaign. There was a great deal of competition for the production of incandescent lamps. Edison was involved in a patent fight with William Sawyer who had carbon filament lamps also. Sawyer's claim and work went back to the 1870s. Another competitor, Elihu Thomson entered the market with a modified Edison lamp. In Europe, the Swan incandescent lamp was challenging Edison both legally and commercially. None of Edison's competitors had the financial backing or full system control of Edison. Edison also had the best publicity machine. In 1883 Edison arc light competitor Brush Electric bought the patent rights for the Swan incandescent lamps. Morgan associates were also active in trying to buy up competing patents. Morgan was not interested in manufacturing, but he wanted financial control of the full system and because of this Edison was able to stay on the DC path. Thomson-Houston Company was, with Westinghouse, one of the first to see the potential of AC power. The incandescent lamp patent battles of the early 1880s distracted Edison and Morgan to some degree from the rising interest in AC power. They, of course, had bought the patent purchase options on the improved AC transformer that had the potential of quickly giving AC current the edge. In Europe a lot of research was looking into AC power, but Edison was fully committed to DC by 1885.

Westinghouse' s Garrison Alley think tank progressed slowly as personalities and egos clashed. Westinghouse was needed often to keep them on task. The work of AC current generation started advancing rapidly at Westinghouse Electric when Stanley became ill and had to return to his home in

Great Barrington, Massachusetts (Berkshire Hills). Great Barrington was in the Berkshire Mountains, not far from Westinghouse's Erskine Park home. Westinghouse sent Belfield to Great Barrington to set up a research plant for the development of AC power. The two engineers hooked up a 25 horsepower Westinghouse steam engine to a Siemens alternating current generator for power generation. They modified and re-engineered the Gaulard-Gibbs transformer to deliver the power. The system applied was that developed by the Westinghouse team at Garrison Alley, but engineered on site by Stanley and Belfield. Great Barrington was a small town consisting of 13 stores, two hotels, two doctor's offices, a telephone exchange and a post office. Stanley and Belfield wired all of these with electric lamps. On March 17, 1886, while Westinghouse was in Pittsburgh, Stanley moved forward to light a single store as an example for the area's businessmen. It was a slight of Westinghouse that would later allow him to claim the title of the first to light with AC and start his own company. He then telegraphed Westinghouse of the success, and Westinghouse came for the larger Saturday night planned demonstration. They started their small power plant to generate high voltage AC current about a mile away on March 20, 1886, and then used the modified transformers to step down the current at the buildings. This was the first large-scale demonstration of AC lighting. Westinghouse immediately telegraphed Frank Pope to sail for Europe and purchase the rights to the Gaulard-Gibbs patents. Meanwhile Westinghouse took the train back to file for a number of system patents related to the success. Stanley would ultimately start his own AC power company in 1890 using the Westinghouse system, but he lost this right in lawsuits. Westinghouse started working on an electrical meter that day as well. A meter would be needed to measure and charge for electrical power, something he had learned from his gas distribution experience. Westinghouse realized that metering was necessary for profitable distribution of energy. Again he was ahead of Edison and Morgan in the development of a true power generation business. Morgan would come to realize too late that Westinghouse would have made the better business partner. The meter project was ultimately turned over to Oliver Shallenberger.

Edison had worried little about metering, and started charging for his electrical power on the per lamp rate. A few years later, Edison's love of chemistry helped him develop a type of chemical meter based on electrolysis. The chemical meter tried to measure amperes (current) used by hour. The Edison meter was a simple jar with two zinc plate electrodes, which the incoming customer current flows through. The plates were taken out every month and weighed. The bill was then related to the weight change, which in theory reflected usage. It was error prone and inaccurate. Westinghouse's business sense helped him go right to developing an accurate meter. Within a few months of his Great Barrington success he had one. Oliver Shallenberger continued to improve on this meter. He patented his meter in 1888 and it became the standard in the industry. This last improvement of the meter happened by accident in the laboratory. This induction meter was actually a form of an AC electric motor, although Shallenberger didn't realize it. A few years later Nikola Tesla would point it out to Shallenberger. Westinghouse demanded the meter read very directly with analog dials, like a water meter. Westinghouse himself would actually patent such a meter for natural gas.

Westinghouse wasted no time in building on his success at Great Barrington. Using a power station, Stanley sent current three miles from downtown Pittsburgh to Lawrenceville and kept 400 lamps lit for several weeks in October of 1886. Westinghouse studied every wire of the setup before calling in the press. This was very significant; Edison's system couldn't go much last a quarter mile. Westinghouse spent many hours watching this system work. Many tests were tried as Westinghouse engineers collected data. In a few more months, Westinghouse had a larger power station lighting Greensburg, Pennsylvania, outside of Pittsburgh. On Thanksgiving of 1886, Buffalo, New York, was lighted by the Westinghouse AC system. Westinghouse had now entered Morgan and Edison territory. Westinghouse Electric was off and running. They needed over three thousand employees in the three years. The construction company of Westinghouse, Church, Kerr, and Company was formed to build power plants. Edison and Morgan were still confident in the DC system, but Westinghouse's successes had to be addressed. Edison expressed the superiority of DC current, as Buffalo, New York, turned to AC in the following memo:

> Do you know that the Dynamo Batch made for transforming is a very perfect contrivance, in fact it is perfection. The more I study our converter business, the more I am satisfied that we shall be able to give Westinghouse all the law he wants on this particular subject or any other. None of his plans worry me in the least; only one thing that disturbs me is the fact Westinghouse is a great man for flooding the country with agents. He is ubiquitous and will form innumerable companies before we know anything about it. When it comes to dollars and cents nothing that anyone else could possibly do could touch us in least. The moment capital has confidence and will furnish unlimited capital, if they can make 8 per cent we will think no more of putting in $20,000 feeders than the Pennsylvania Railroad does of spending $100,000 to straighten a curve.[38]

The fact was the Westinghouse AC system was cheaper and delivered lower cost electricity. Westinghouse continued to win municipal contracts based on his cost advantage. Edison had however, the support of Morgan's money trust and he had locked up the big cities such as New York, Chicago, and Detroit. Particularly, when Westinghouse tried to move into New York City, the politicians closed ranks on him. He was forced to use overhead wiring while Edison laid underground wires. The overhead caused serious safety problems and resulting accidents got charged against Westinghouse in the press. Westinghouse's AC system seemed to be favored by most of the world's electrical engineers, but Edison's reputation countered this. In

38 Robert Conot, *A Streak of Luck: The Life and Legend of Thomas Alva Edison* (New York: Simon and Schuster, 1979), 254

theory, AC current could supply from hundreds of miles away. Even Edison's three-wire feeder system was limited to a few miles. In fact, most Edison installations had the power station within a few yards. Edison had several advantages in a superior incandescent bulb, improved safety, and an electric motor. Edison played every card and remained ahead of Westinghouse. In1890 Edison had 870 central stations in big cities, while Westinghouse had power 300 stations in small and mid-size cities. Westinghouse Electric's growth was extraordinary, with sales quadrupling from $800,000 in 1887 to $3 million in 1888. Westinghouse was always on the offense while Edison was playing defense with few exceptions. Edison did invest heavily in a "pro-magnetic" generator that would require no Westinghouse steam engines, but even the great wizard couldn't make this futuristic approach practical. Westinghouse had launched projects to address his two weaknesses — a lack of an electric motor and an incandescent bulb. Edison, however, played to the safety factor.

The first ten months of Westinghouse Electric were indicative of the different nature of the electrical business. It was a business that required huge capital to implement projects. Westinghouse Machine Company and Westinghouse Air Brake Company were manufacturing cash cows, but Westinghouse Electric was often cash short. It was formed with 20,000 shares of stock in January of 1886. Westinghouse originally had 18,000 shares with the other 2000 being distributed among seven partners. By October 1886 Westinghouse was down to 9,605 shares after almost monthly sales to generate needed cash. This type of stock selling and bond issues would be the way of life for Westinghouse Electric. This was where Westinghouse's bias against bankers hindered him. He needed to establish a banking relationship, but he mistakenly felt he could finance the company. It was typical of companies of this period to have bankers on the board of directors, but Westinghouse felt the board should consist of operating people. This lack of a banking relationship would always be one of his weak points, and ultimately, would lead to the down fall of Westinghouse Electric.

In late 1886 Westinghouse moved to correct his weakness in incandescent lighting by purchasing the rights of the competing Sawyer. Westinghouse had beaten Morgan and Edison to these rights as they were tied up in a court battle over them. Edison had suggested early on to Henry Villard to merge with United States Electric Company who owned the rights to the Sawyer patent. Edison realized the technical superiority of the Sawyer incandescent bulb. The Sawyer patent was a major challenge to Edison, but while sound technically, Sawyer, an alcoholic and trickster, had lied often in his patent application. These lies testified to by Sawyer's brother would ultimately win the court case for Edison. In the short run, however, the Sawyer bulb was much improved over Edison's and easier to manufacture. Meantime when Westinghouse purchased the rights, he was drawn into the court fight. Westinghouse wrote Edison to come to Pittsburgh and talk, offering to show him all of his works. Edison refused suggesting Westinghouse was crazy.

Edison Electric and Thomson-Houston were the two major competitors. Westinghouse Electric in 1887 didn't really pose a threat to either. Westinghouse wanted to be a full system player in the electrical lighting market. He had at least a rudimentary AC generator and transformer. Westinghouse had some patents and expertise in arc lighting, but Edison and Thomson-Houston controlled the incandescent lamp patents. Morgan had already advised Edison of Westinghouse's potential danger, so Westinghouse was forced to buy up all the non-Edison patents for the incandescent lamp. Before Thomson-Houston fully appreciated the threat from Westinghouse, he purchased the rights from them for the Sawyer-Man incandescent lamp. It is an interesting side note that William Sawyer was one of the few eccentric electricians. Westinghouse didn't have in his Garrison Alley research operation. Sawyer would have surely fit the mold, having an erratic personality and the inability to work with others. Sawyer would eventually be sentenced to four years of hard labor for shooting a fellow worker.

Under Morgan's radar screen was the fact that Westinghouse was building a massive engineering organization. Edison's Menlo Park was in many aspects a one-man operation with many hired hands. Westinghouse was hiring the world's best engineers, buying their patents, and developing newly graduated electrical engineers. Edison would do none of these things. Edison passionately disliked college educated scientists and engineers. Edison hired people to do his calculations and throw a little more technical jargon in his patent applications, but the creativity of Menlo Park was Edison's brain. Westinghouse's company, however, had a corporate creativity of many engineers and scientists. Westinghouse encouraged their creativity, and he had the rare ability to develop the theoretical ideas of these young geniuses and produce models and drawings. His dinner meetings at Solitude with his engineers can be considered one of the earliest examples of a think tank. One necessary piece of apparatus for Westinghouse 1886 entry into power generation was the voltage regulator. The invention of this device came from Louis Stillwell, a recent college graduate in Westinghouse's employ. Edison, on the other hand, had the winning chip in 1884 but didn't realize it. With Stillwell, Stanley, Pope, Belfield, Pantaleoni, and Shallenberger, Westinghouse had assembled the world's greatest AC electricians in his organization, save one. In Edison's employ was Nikola Tesla, the inventor of the AC induction motor.

Nikola Tesla seemed to fit the mad scientist in Mary Shelly's *Frankenstein*; he even looked the part, tall (6' 6"), dark, and handsome. Nikola Tesla was a brilliant, eccentric, and theoretical genius, all things Edison despised. Even more so Tesla loved publicity and high living, and publicity was something Edison could not share well. Born in 1856 in Smiljan, Croatia, he was a self-trained scientist, although he had some college training. Tesla had made a name for himself in European circles. He also had demonstrated electrical skills that would be applied. In the early 1880s, he was one of the young scientists interested in AC current. Tesla invented an AC electric motor in 1882, but had no practical AC power source to promote its use. He moved to Paris to take a job with French Edison Company in1883. Edison had been interested in Tesla's work on electrical motors, and he hoped to adapt his work to DC motor applications. Henry Villard, Morgan's agent in Europe hoped to bring Tesla in for AC current research. When Tesla came to America in 1884, he clashed immediately with Edison who would have nothing to do with AC current, and had a growing dislike for Tesla personally. Tesla's ability to deal with extreme abstractions such as the different phases of alternating cycles further upset Edison. Edison never could come to terms with the theoretical complexity of AC current, and engineers, like Tesla threatened him.

While Westinghouse built an organization for the battle, Edison looked to his publicity network. Edison just would not give in to AC power because of its complexity. He understood the simplicity of DC current and had grown up with it. It didn't require the mathematics and science theory needed to apply it. Edison had a near phobia for math. DC was the ideal power source for tinkers like Edison. AC current was tough to visualize with its phases, frequencies, and cycles. It was far from the straightforward chemical principles that Edison had shown genius in applying. Furthermore, AC current didn't lend it self to the trial and error approach of Edison. Westinghouse actually had a "calculating department" with graduate electrical engineers, such as Benjamin Lamme. AC current work required one to move from theory to practice, and required an understanding of calculus, which only graduate engineers, would have any experience in. The search for the incandescent filament was Edison's Magnum opus. It was chemical in nature and could only be resolved by trial and error. Even in the midst of the war of the currents, Edison lacked the staying power for long-term product development and research. In asking Morgan in 1888 for money to build a new invention factory at West Orange, New Jersey, this was apparent:

> My ambition is to build up a great industrial works in the Orange Valley, starting in a small way & gradually working up: the laboratory supplying the perfected invention models, patterns, and fitting up necessary special machinery in the factory for each invention. My plan contemplates to working only that class of inventions which require but small investment and of a highly profitable nature, and also of that character that the articles are only sold to jobbers, dealers, etc. No cumbersome invention like the Electric Light. Such a works in time would be running on thirty or forty special things of so diversified nature that the average profits would scarcely ever be varied by competition, etc.

Morgan and his partners basically controlled Edison at this point and they needed him in the fight with Westinghouse, not off inventing little gadgets. Morgan was also pouring money into Edison's DC system by forming a copper syndicate and even had agents in the African Congo trying to control the rubber market needed for wire insulation. Edison had already spent ten years on incandescent lighting and he wanted to move on. Edison may well have been what today we call burned out. One commentator noted, "In 1879

Edison was a bold and courageous innovator. In 1889 he was a cautious and conservative defender of the status quo."

On the other hand, Westinghouse seemed to be entering years of peak performance. Nikola Tesla in 1914 looked back to 1888 and described Westinghouse:

> Though past forty then, Westinghouse still had the enthusiasm of youth. Always smiling, affable, and polite, he stood in marked contrast to the rough and ready men I met. Not one word which would have been objectionable, not a gesture which might have offended —one could imagine him as moving in the atmosphere of court, so perfect was his bearing in manner and speech. And yet no fiercer adversary than Westinghouse could have been found when he was aroused. An athlete in ordinary life, he transformed into a giant when confronted with difficulties which seemed insurmountable. He enjoyed the struggle and never lost confidence. When others would give up in despair he triumphed.[39]

Westinghouse had shown a love for technical trials and debates with the fight of braking systems. Westinghouse was always confident in his ultimate victory. Like a good football coach, he was flexible and adaptable to the developing situation. More importantly, the battle or competition was not personal but winning was. Westinghouse prepared for victory. In the long run neither Westinghouse nor Edison would win personally. Morgan would ultimately have both companies and see Westinghouse and Edison sitting on the sidelines. The simple fact appears that Edison was not interested in the business end of the battle for the currents. Edison loved competition too, but for him it was personal. He would prove inflexible as well.

By 1888 Edison and Westinghouse were finding themselves in bidding battles for municipalities across the nation. The copper market was rising rapidly and threatened to take Edison out of the competition. Westinghouse had a major advantage until Morgan got control of the copper market. In 1888, however, a non-Morgan syndicate got a hold of the copper, which sent fear to both Morgan and Edison. The following from the March 1888, *Journal of Engineering* highlights the story:

All the electrolytic copper in this country is now firmly in the grasp of the syndicate. There appears, in fact, nothing to prevent prices from being advanced to any figure the syndicate may wish. This unfortunate and ominous turn of events was a real blow to the Edison Electric Light Company. For instance, in the spring of 1887, the company had been putting together a bid for a Minneapolis central station powering 21,700 lights. They estimated the feeders at 254,000 pounds of copper and the main at 51,680 pounds. At seventeen cents a pound, copper costs would total $51,965. Each one-cent rise in copper pushed the costs up $3,056. Three-cent rise —for copper prices were escalating steadily —would add $9000 to the almost $52,000 price tag for copper. In painful contrast, the Westinghouse AC central plants required a third as much copper.

39 Nikola Tesla, "Tribute to George Westinghouse", *Electrical World & Engineer*, March 21, 1914

Morgan could see a company beyond Edison now. Morgan wasn't interested in winning the technical battle over currents; he wanted control of the electrical market. This type of fierce competition was what Morgan always maintained was wasteful to the stockholders of everyone involved. He would help Edison fight the technical battle, but he would also look at gaining control of the competition. This strategy would lead to the ultimate merging of companies into General Electric.

Morgan had now realized he needed to take management control of Edison's electrical companies. Edison was locked in patent battles, a stubborn commitment to DC, resentments toward key people in the industry, an inability to compromise, a wandering research effort, and mismanagement of the operating companies and departments. Morgan men inside the organization reported that at Edison's largest manufacturing plant at Schenectady, "workers were literally standing on one another's feet." It was these types of inefficiencies and mismanagement that Morgan based the need for syndicates and trusts on. A lot of this took place as Morgan's man Henry Villard had been in Europe. Edison was constantly getting in the way of Villard to implement improved processes and organization. Edison refused to join in on discussions of mergers. In particular, Edison had personal problems with the management of Thomson-Houston. Furthermore, Edison had banded all research on AC current. Edison's company was failing profit wise through 1884 to 1889, and Villard allowed Morgan money and debt obligations to increase. In addition, Edison himself was becoming cash poor, often short money for personal expenses, which would play into Morgan's hands. Edison started to fear bankruptcy, which was now a real possibility. Finally, in 1889, Villard succeeded in forcing Edison out of an active role and out of any real ownership forming Edison General Electric. Villard, through Morgan's banking syndicate teamed up Deutsche Bank and Siemens Electric to form a type of world cartel. Villard bought out his major competitor, Thomson-Houston a year later. The new General Electric probably controlled 75% of the market at this point in 1890 with Westinghouse being the only major competitor. Still, AC current's potential had to be addressed. Edison was now reduced to the role of attack dog.

Westinghouse had reorganized his resources in 1888 for a goal of future dominance in the electrical industry. His Garrison Alley test operation was now a world-class research center. It was divided in a DC/AC test department, an incandescent lamp department, and an AC test department. In 1889 a motor test department was added. The motor test room spearheaded a new goal of Westinghouse — electrical railways. Albert Schmid was the Superintendent of the Garrison Alley center, and Westinghouse was a daily visitor. Westinghouse spent as much time with the organization as he did with the electrical equipment. When Westinghouse moved into a new area, he went out and hired the necessary engineering expertise to support it. Besides gathering electrical engineers from Europe and competitors, he was bringing in graduate engineers from colleges. Benjamin Garver Lamme would be the first of these in1889. Lamme was a graduate of Ohio State University. The newly developed Westinghouse system required him to start his career at the testing rooms. Edison and most of the industry were prejudiced against educated electrical engineers; they wanted practical electricians. Westinghouse realized that AC current needed the scientific man trained in its abstractions. Lamme was typical of the graduates. He entered the testing rooms on a salary of $30 a month. The work was far from glamorous consisting of polishing brass work and oiling machines. Schmid and the foremen watched for opportunities to challenge the youth. Electrical calculations became more and more important and this was one place the college education paid off. Lamme moved fast through many calculating assignments as his salary rose to $70 a month. After a year or two the young engineer would be rotated into "road work" in the field or at the manufacturing plant. Westinghouse at this point was far from a lone inventor, in fact, inventor hardly applied. Westinghouse was a manager and developer of inventors, and that would be the seed of his success. Westinghouse still had his "pet" projects, on which he lavished his time and interest, but he allowed an amazing amount of control to his engineers. In many ways Westinghouse was the antithesis of Edison.

Any researcher will find difficulty in trying to fully understand what drove Westinghouse. He seems at times too pure to be real. He wanted to win, but did not want the trophy. We can better understand men like Edison, Tesla, Stanley, Sawyer, and Thomson. Westinghouse certainly didn't fit the robber Barron image, so popular during the Gilded Age. He had the humility of a monk, which doesn't fit his zealous drive to win. He cared little for publicity, headlines, and his name on anything other than the company. He gave more credit than he needed to. His philanthropy was always behind the scenes. Like his tombstone, he played down his legacy. He did enjoy the good life, but respected the needs of others. He had great power, but rarely used it. As you research his almost too pure image, you have to return to his upbringing and Puritan roots. His family and social behavior demonstrated the same moral consistency. He demonstrated tolerance that few managers were capable of with some very difficult people. He more than any other industrialist fully embodied the "Protestant Work Ethic," including the personal enjoyment of the rewards.

Meanwhile, Tesla was splitting more and more with Edison. He was lecturing around the country on his AC motor and the potential of AC power, which infuriated Edison. Tesla broke off from Edison in 1888 forming Tesla Electric, whose main asset besides Tesla himself were his AC motor patents. The Edison and Westinghouse people were lost in the complexity of Tesla's electrical theory, but the Westinghouse people showed more vision. William Stanley who felt Tesla offered little over his discoveries had dampened Westinghouse's initial enthusiasm. Westinghouse's lawyer, E. Kerr, felt Tesla's patents and studies needed to be evaluated thoroughly. Westinghouse sent Henry Byllesby (then vice president of Westinghouse Electric) to

Tesla's Liberty Street lab in New York. Byllesby reported the following back to Westinghouse:

> His description was not of a nature, which I was enabled, entirely, to comprehend. However, I saw several points, which I think, are of interest. In the first place, as near as I can get at it, the underlying principle of this motor is the principle which Mr. Shallenberger is at work on at the present moment. The motors, as far as I could judge . . . are a success. They start from rest, and the reversion the direction of rotation is suddenly accomplished without any short-circuiting . . . In order to avoid giving the impression that the matter was one which excited my curiosity, I made the visit short.[40]

Westinghouse reviewed everything Tesla had written and was able to see through the abstractions where others failed. Westinghouse wrote Kerr, "If the Tesla patents are broad enough to control the alternating motor business, then Westinghouse Electric Company cannot afford to have others own the patents." Westinghouse remembered how just a few years earlier Edison had blocked him with the patent rights on the Z.B.D. transformer. Westinghouse also needed an AC motor eventually to be fully competitive with Edison's DC system. Also, Westinghouse had already made a decision to move into electric railways, which needed an AC motor. He would prefer to use AC to augment his power generation business, and a Tesla AC motor could theoretically drive streetcars.

In July 1889, Westinghouse called Tesla to Pittsburgh to finish the negotiations. Tesla and Westinghouse hit it off. Tesla arrived in Pittsburgh in the middle of a high-pressure heat wave making the smoke, dirt, and heat combination unbearable. Westinghouse had Tesla come to Solitude to escape the heat. Solitude was a white brick mansion with its awnings, and in the shaded neighborhood of Homewood offered some relief, and Tesla was an excellent and charming dinner quest, with whom Westinghouse would spend hours in profitable conversations. Tesla was an ideal companion for Westinghouse, a great dinner conversationalist, a billiards player, and an early morning companion to discuss science and engineering with. They could communicate on the complexities of AC current and had a mutual respect on a personal level. The agreement that they came to would secure the future of Westinghouse's ultimate victory over DC systems. Westinghouse paid a combination of cash, stock, and future royalties. Money was never the issue for either men, who would spend their last dime on a new piece of technology, but they both had a deep respect for intellectual property rights. Over a hundred years later as part of an advertising campaign, Westinghouse Electric took out full-page ads with pictures of Westinghouse and Tesla calling it "the perfect partnership." Indeed it was just that. Tesla's strength was theory, but often he failed to translate it to the grasp of practical men whereas Westinghouse's talents were in adopting scientific principles to practical use. The deal was at estimated around $500,000, to be paid in installments. The deal included

40 Marc Seifer, *Wizard: The Life and Times of Nikola Tesla* (New York: Citadel Press, 1998), 49

minimum royalty payments for use of the patents. It was a very good deal for Tesla. Westinghouse in return got the patent for the AC induction motor and various other devices. Tesla would also spend a year in the testing rooms of Westinghouse Electric working on the motor. It did cause a rift in the Westinghouse organization. William Stanley would leave to pursue the development of his own polyphase AC motor. Shallenberger seems to have resented Tesla's favored position with Westinghouse as well, but Schmid quickly befriended Tesla and coauthored a number of patents with him.

Westinghouse did his best to help Tesla fit into the organization, which already had its share of egoist people. Tesla, however, was used to not fitting in or being completely understood. More importantly, he had Westinghouse's support and the technical freedom he had lacked in Edison's organization. Westinghouse was an expert at managing temperamental scientists. The problem became Tesla's love of high society that Pittsburgh lacked. Westinghouse was able to set him up at Pittsburgh's Monongahela House, which could rival most New York hotels. The Monongahela House could boast visits of many presidents starting with Abe Lincoln, as well as, as British royalty. The Monongahela House's Blue Points, boiled Kennebec Salmon, Sweetbreads Glace a la Financiere, Breast of Pheasant a la Victoria, and Plum Pudding (from a 1897 Monongahela House menu) were as good as any, but Pittsburgh lacked the social circles of New York. High society was Tesla's passion and weakness, and would ultimately shorten his work in Westinghouse's organization. Still, the months spent in Pittsburgh would give Westinghouse Electric a competitive advantage for years.

Tesla's polyphase generator was one of his complex electrical designs. The alternating current did just that; it alternated voltage between peaks and valleys of the cycle. DC current delivered a constant level of power and voltage. The alternating current created lags and surges in power, which could lead to dead spots in motor operation. Tesla overcame this by generating three different cycles out of phase with each other. This polyphase approach allowed for one phase to always be a peak, which in a way simulated constant DC power. The actual timing of the cycles was referred to as a phase. These out of phase cycles would eliminate the dead spots in motor operation. Competing AC current generators were single phase. Westinghouse and his engineers saw the brilliance in Tesla's solution of polyphase. There was a problem, however. Tesla's polyphase system was based of a generating frequency of 60 cycles per minute (Hertz). All of Westinghouse 's systems generated at 133 cycles per second. In particular, Westinghouse's modified Gaulard-Gibbs transformer operated at 133 Hertz as well as Shallenberger's meter. Shallenberger felt that Tesla should work on adapting his motor to the 133 frequency with 120 power stations already setup at 133 Hertz. Tesla tried but failed and ultimately believed 60 Hertz was necessary. Tesla was also a poor team player and work stalled. Westinghouse became indecisive with such a split among his best technical minds. Westinghouse favored in the face of a deadlock pushing Tesla to adapt his motor to 133 Hertz. The internal struggle continued for almost three years. Finally, Westinghouse himself stepped in, seeing the error of the 133 frequency, and decided to convert the system rather than Tesla's motor. One engineer that helped Westinghouse see the light was the young Benjamin Lamme who had the technical grasp of polyphase AC current. 60 Hertz (cycles) became the AC standard and is still used today. It would lead to the invention of the Tesla coil. Still, it wasn't until 1891 that Westinghouse Electric started to make the necessary conversions. Tesla, however, left not being a team player. He would always have high praise for Westinghouse, and Tesla played a key role in the war of the currents.

Westinghouse was one of the few that could converse with Tesla and translate his ideas into practice. The argument for lower frequency become overwhelming as Tesla continued to press the matter with Westinghouse and gained the support of Benjamin Lamme. The resistance of Schmid and Shallenberger came from one weakness of lower frequencies, such as 60 cycle, in producing slower AC motor speeds. Another negative for 60 cycles was that Herman Westinghouse's high-speed steam engine produced by Westinghouse Machine favored 133 cycles AC generation. The high-speed steam engine of Westinghouse generated a significant amount of business as a subcontractor to Thomson-Houston 133 cycle generators. While Westinghouse probably realized the advantage of 60 cycles, but for business reasons wanted Tesla to adapt his motor to 133 cycles. On the plus side, lower frequency reduced the cost of generator construction, improved the distance AC can be transmitted, and allowed for more efficient conversion to DC using Lamme's rotary converter. Westinghouse asked his lab to test 60 cycles for use in lighting systems, which that found satisfactory. Westinghouse then ordered all designs to be built on a 60-cycle design. The decision left hard feelings among the Garrison Alley research group and further distanced Tesla from his associates there. Tesla stayed only about a year in Pittsburgh, living at Westinghouse's expense in Pittsburgh's best hotels. Tesla always wanted to get back to New York preferring the social life, better hotels, and his favorite restaurant —Delmonico's. Delmonico's had become his "home"; he loved to eat, entertain there, and play pool. Tesla had also broken into New York's social elite known as "The New York 400."

Probably the most surprising relationship was the lifelong one between Westinghouse and Tesla. Tesla, while outgoing, had tended to not work well with other engineers. He could be very hardheaded and uncompromising, yet he was a tall man with big ideas just like Westinghouse (both were over six foot two inches). Both these features seem to have appealed to Westinghouse. One of Westinghouse's most famous engineers Benjamin Lamme described Westinghouse in his autobiography:

> Mr. Westinghouse seemed particularly attracted by men of large stature, apparently taking such people into his confidence much more readily than men of smaller build. I do not know whether this made much difference, in the long run; but apparently some of us "smaller built" men had to

> attract his attention almost entirely through our work, rather than our personalities. Also, if a man talked "big" enough, he would attract Mr. Westinghouse, to a certain point at first; but if he could not live up to his talk, Mr. Westinghouse would soon drop him. [41]

The war of the currents was a complex mixture of AC and DC power usage and distribution, incandescent versus arc lighting, electrical motor technologies, huge egos, and financial manipulation. Many boil it down to a technical struggle and debate, but from a technical standpoint, AC after Westinghouse's efficient generators and Tesla's transformers, was clearly superior. The complex nature of electrical current allowed the war to continue for years, but technology and economics favored AC in the long run. Westinghouse was winning the technical battle easily, since his engineers dominated the technical societies and journals of the day. A Westinghouse electrical engineer and officer had even risen to editor of the *Electrical Engineer*. The status of electrical power generation at the end of 1887 told the story. Edison had 121 central power (DC) throughout the United States, but primarily in densely populated areas that lowered its distribution costs. Westinghouse in about a year of market operation had 68 AC central operating or under construction. Thomson-Houston had 22 AC stations in operation using Westinghouse transformers and steam engines. Even Edison used Westinghouse engines at many locations. Thomson-Houston was unique in that they were using primarily arc lighting, which was their forte. The Edison system required about four times as much copper wire, and the price of copper had gone from 10 cents a pound to sixteen a pound by the end of 1887. Financial men within the Edison organization were starting to pressure Edison to research AC current, but Edison would have none of it. By the end of 1887 Edison had started to lose bids to even big cities like Minneapolis because of the rising cost of copper.

Edison could not refute AC current on a technical basis, but he correctly saw the safety problem with AC current. Early on, Westinghouse systems were making headlines because of the danger of high voltage AC current. One of the earliest deaths was a horse that stepped on a AC line in Buffalo; an ancillary result was the death of the rider. There was a house fire in Pittsburgh also that highlighted a related but different hazard. A boy was killed by touching a line and a Westinghouse repairman also died. Morgan controlled most of the big city presses, and these stories started to make headlines across the country. Even while AC current got the praise of most engineers, they did not deny the danger of alternating current. Edison and Morgan both however felt they could exploit this weakness of AC current. AC certainly had its problems. In New York City AC was delivered by overhead wires, which added to the danger. The Edison DC system had been done with underground cables due to the political support of Edison and Morgan to obtain right of way. The AC overhead wires caused very deadly

41 Benjamin Garver Lamme, *Benjamin Garver Lamme: An Autobiography* (New York: G. P. Putman's Sons, 1926), 116

problems as they still can today. The rapid growth of some areas of the city had changed the environment into a jungle of overhead wires. Editorials raged against these AC wires as an eyesore. Another problem became obvious with the first big blizzard of the electrical era —the Blizzard of 1888. The blizzard roared in with ice followed by twenty-four inches of snow, and then the characteristic east wind with subzero temperatures. The blizzard left live wires everywhere in a web of tangled wires. Edison underground wires were the best for either AC or DC, but the cost was prohibitive with easy access for maintenance and expansion. "Telephone" poles were carrying telegraph, telephone, and power lines at a fraction of the cost of burying lines. Amazingly, Westinghouse had been an early supporter of buried lines, something that would not be fully implemented until a hundred years later. What followed were years and millions of dollars spent in advertising and publicity campaigns on the overall dangers of AC by the Edison organization. It forced Westinghouse to hire a publicity executive to handle that part of the business.

In 1888 Edison would find a strange ally in his campaign against AC current. Harold P. Brown was an electrical engineer at the Columbia University. Brown was a crusader against AC current driven by the death of several colleagues. Brown had compiled a list of over 60 deaths in a two-year period. As part of his campaign against AC current, he started to promote the killing power of AC. Experimenting with execution of dogs and cats, Brown started to talk about its possible use for killing criminals. At the same time the Society of Prevention for Cruelty of Animals had asked Edison if electricity offered a humane way of killing animals. Edison and Brown became the perfect marriage. Brown was given free reign in Edison's lab to perform experiments. Edison often called in reporters to witness these trials. He paid twenty-five cents a piece to local children to bring in cats and dogs. They promoted the term "Westinghoused" for electrical execution.

Westinghouse had early on tried to build a better relationship with Edison, with no success. Public fights, lawsuits, and name-calling were inconsistent with the character of George Westinghouse. Westinghouse had met Edison a few years earlier during a tour of Menlo Park during Edison's public light ups. Personally Westinghouse found Edison a very likable man, and Edison was a courteous host, but that was before Westinghouse had fully entered the business. Before the formation of Westinghouse Electric, Westinghouse had even pursued Edison in hopes of putting a DC power plant at his home Solitude. On June 7, 1888, Westinghouse wrote Edison directly, "It would be a pleasure to me if you find it convenient to make a visit here in Pittsburgh when I will be glad to reciprocate the attention shown me by you."[42] Edison quickly and courteously replied that he was busy at the lab. Once Westinghouse Electric was formed, all things changed. Edison disliked

42 Letter from George Westinghouse to Thomas Edison dated June 7, 1888. See Thomas Edison Archives website, Rutgers University

Westinghouse because he threatened his electrical prominence, and Edison loved a good resentment.

Westinghouse even wrote a letter to the *New York Times* protesting the publicity campaign. Westinghouse did not even realize that Morgan and his associates were controlling the debt for the *New York Times.* Morgan's influence certainly helped in the success of this campaign of Edison's. The *New York Times* and other New York papers were relentless in their reporting of every AC current accident, as well as many editorials. Edison also was letting the debate become personal as he had with his earlier competitor —Thomson. Westinghouse did not enter into a personal battle; he always tried to keep it in the technical arena. Tesla would years later state that only Westinghouse could have taken on Edison, Morgan, Brown, and the *New York Times* over AC current. Westinghouse saw some success in addressing the "electric wizard" in print, and started to reply to all of Edison's letters to the press and journals. Westinghouse even wrote a ten-page reply to one of Edison's letters to the North American Review published by the University of Northern Iowa in 1889 (the full reply is enclosed in the appendix because of its historical significance). Westinghouse was never a journalist, but he was an excellent writer capable of forming and supporting a thesis. Westinghouse presented the following statistics in that reply deaths in the city of New York show that there were killed by street-cars during the year 1888, 64 persons; and by illuminating gas, 23; by omnibus and wagons, 55; making the number killed by the electric current (5) insignificant when compared to other causes . . . and they may all be prevented by the employment of reasonable and well understood safeguards."[43] Westinghouse showed amazing persistence against an Edison team that was said to have put "persistence" in the dictionary.

The Edison/Brown demonstrations became even more dramatic. A horse and cow were electrocuted for the press. At one point Brown was searching for a circus elephant to electrocute. Brown even went to the Times to challenge Westinghouse to an electrical duel running the two currents through their bodies. As the state of New York considered AC for executions, Brown devised a cap and shoe to do the job, using Westinghouse AC dynamos. Westinghouse failed to stop the use of his generators, but watched to prevent such future applications. In August 1890, William Kemmler, a convicted murderer, was electrocuted. The system bungled the job turning it into a cruel spectacle. Westinghouse said, "They could have done better with an axe." Still, the overall campaign had been extremely successful in putting fear in the general public. The White House had been wired with an AC system, but President Harrison and his wife were afraid to use the switches. Westinghouse even stopped work on his AC motor because of the public hostility to Tesla. Edison, Brown, and Morgan had taken the opportunity to "repay" Tesla for his defection to Westinghouse. One positive part of the

43 "Reply to Mr. Edison's letter," *North American Review*, December, 1889

campaign was that it forced Westinghouse to move to the 60-Hertz (cycles) frequency, which was considerably safer than the 133-Hertz system.

At the end of the 1880s, Henry Villard, a Morgan man, was running Edison General Electric. He shared Morgan view of an "electrical trust." Basically in Europe, Morgan and Villard had a trust agreement in place with Siemens. In the United States the situation was far different. Three big companies, Edison General Electric, Thomson-Houston, and Westinghouse Electric were locked in fierce battle. There were also a handful of key smaller companies that owned critical patents or options on patents. It was hyper competition resulting in patent battles, price-cutting, negative publicity campaigns, technical battles, spying, personal fights, and employee raids. For J. P. Morgan this was capitalism's worst nightmare, and in this case Morgan had a real point. Huge amounts of money were being wasted in legal battles alone, and that might have been only the tip of the iceberg. All three companies were in a patent buying frenzy as a defensive strategy. The big three companies had also built up technical firepower. General Electric had Edison, Westinghouse had Tesla and an army of young engineers, and Thomson-Houston had just hired mathematical genius, Charles Steinmetz. Westinghouse was the logical first step in any cartel-forming merger. Morgan had sent a personal agent to talk with Edison about a Westinghouse merger, but Edison would not hear of it. Edison told Villard, he would have nothing to do with a Westinghouse merger or any other:

> You may see things differently from what I do; you may see things through a telescope, while I see the subject through a microscope; still I am sure that if you enter into the slightest connection with him [Westinghouse], it will be at the General Company's expense. We must all expect competition; if not from one person, then from another; but no one can ever convince me that a competitor whose system gives an average efficiency of only 47 per cent can ever prove a permanent competitor for large installations in cities against a system giving 79 to 80 per cent efficiency. But if for other reasons I am incorrect, then it is very clear my usefulness is gone.[44]

Of course, Edison had overlooked the costs of his system, which Morgan and Villard had not. Villard failed as well in his attempts persuade Edison to look at alternating current. Edison's response to the pressing of Villard, "The use of alternating current instead of direct current is unworthy of practical men." The decision in Morgan quarters became clear —Edison had to go.

The Morgan/Villard strategy was to build an international cartel controlled through Morgan money. They needed to bring at least one of either Westinghouse Electric or Thomson-Houston into the fold. Thomson-Houston had been out of the question when Edison was in charge. The plan was now under way to force Edison out, and in the long run that could be done with a merger, which would further dilute Edison's small share. Still, Edison posed an immediate problem that would have to be worked around. Edison had a personal hatred for Professor Thomson, and he was rapidly develop-

44 Conot, 293

ing the same for George Westinghouse. Westinghouse had initially written Edison about coming to a compromise over their costly patent fights. Westinghouse like Morgan saw that millions being spent on lawsuits was extremely wasteful. Lawsuits to Westinghouse were also repugnant and his Puritan catechism stated, *"restitution of goods unlawfully detained from the right owners thereof avoiding unnecessary lawsuits."* Westinghouse was frustrated by the accelerating lawsuits throughout the industry, but he always respected the intellectual rights of the inventor. Morgan was not interested in crushing the competition but wanted control of them. Villard turned to a secret project to develop a world cartel for Morgan. In 1888 without Edison's knowledge, Henry Villard had started making personal trips to Pittsburgh to talk with George Westinghouse. While Westinghouse was extremely suspicious of Villard, he was an ex-railroad executive and that seemed to be the key to Westinghouse's door. Villard had formed the Oregon railway & Navigation Company. Villard had even been given the honor of driving the golden spike at the east west joining of the Union Pacific. He was a gregarious and charming man, the type Westinghouse so loved at his dinner parties. Many have said that Villard was even more driven to form an international cartel of electrical companies than Morgan. In any case, Villard was not interested in technical battles or company battles; he wanted a cooperative trust of companies with control of prices and profits.

Initial talks centered on a compromise on the patent battles, which had cost both companies over $4 million dollars. Villard was joined in the talks by Edward Adams, the second biggest stockholder in Edison Electric and a Morgan associate. Adams tied to talk to Edison about a patent compromise, but Edison would have no part of it. A number of these meetings were at Solitude, out of site of reporters and employees. These meetings only came to light many years later in the notes of Nikola Tesla and famous business reporter Clarence Barron. Westinghouse was in a brief position of strength having no debt or obligations to any banking institution, in addition, cash was pouring in from Westinghouse Air Brake, Philadelphia Gas Company, Westinghouse Machine, and Westinghouse Electric. Edison General Electric had him blocked in patent fights, and Edison's fear campaign was hurting. His brother had also advised Westinghouse that Westinghouse Electric was facing a short-term cash crisis. Villard and Westinghouse realized that Edison would never agree, even though his approval was not necessary.[45] Villard argued effectively that Edison General Electric and Westinghouse Electric were spending millions and losing years, as Thomson-Houston got rich. Villard offered Westinghouse the presidency if they could merge. Westinghouse wouldn't bring himself to bring bankers into control of his operation; he didn't want to go the way of Thomas Edison. Westinghouse realized that through such proposed trusts were not illegal in themselves, they lent themselves to price control by the banking institutions. Just as problematic

45 Harold Passer, *The Electrical Manufacturers: 1875-1900* (Cambridge: Harvard University Press, 1953)

for Westinghouse was the ethics involved in these monopolistic trusts that were being proposed. These core beliefs were perplexing for Villard, since such a trust was legal or at least acceptable under the current administration. He tried to argue Morgan's point that it was for everyone's welfare. These meetings went on from 1888 to 1890 with Westinghouse holding out. Morgan in the end had no leverage, and Westinghouse was confident in the final victory. Morgan, however, knew that Westinghouse was trying to get loans from Pittsburgh bankers to raise cash to cover his research and development efforts.

Next Villard started to make direct appeals to Tesla. Tesla proved as honorable as Westinghouse, and he refused to open the back door to his AC current patents. He did agree to talk to Westinghouse in 1890. Westinghouse became adamant that he would not join such a cartel. Tesla finally ended negotiations in 1892 with the following letter:

> I have approached Mr. Westinghouse in a number of ways and endeavored to get to an understanding . . . the results have not been very promising . . . Realizing this, and also considering carefully the chances and probabilities of success, I have concluded that I cannot associate myself with the undertaking you contemplate.[46]

Two ethical men and a hardheaded inventor in Edison had foiled Morgan, but in 1891 he turned to Thomson-Houston as well. Henry Villard was sent in February of 1891 on a clandestine mission to see Charles Coffin, president of Thomson-Houston, as well as the factory. Coffin was a good employer and considered a fair man, although, Westinghouse never liked or trusted him. Coffin was also a forceful leader who would have to be satisfied. He demanded equality in the merger and recognition of certain methodology. In reality, the capitalization and assets of the two companies were close. Thomson-Houston had been well known for its factory system, which would aid the disarray found in many Edison General Electric operations. Morgan cared little for organization structure initially as long as he had strategic control in the marketplace. In fact, Morgan wanted a smooth merger; he had no interest in a wasteful takeover battle or internal struggles. Coffin was named president of the new company to be called General Electric, Edison name being dropped to send both Edison and Thomson shareholders the same message. Edison had to go, in Morgan and Villard's minds; he was diluting the research focus with the phonograph and other projects. His hardheadedness had put the company behind in applying AC current. In addition, his personal resentments of Thomson, Westinghouse, and Tesla were getting in the way. Finally, Edison's total resistance to alternating current in any way was becoming a major problem. Edison remained on the board but was a face man only reduced of any operating say. Most of the Board was made of Morgan men, with Thomson-Houston assuming key operating positions. Edison attended only one meeting, finding that he had no say in

46 Nikola Tesla to Charles Villard, October 10, 1892 [Houghton Library, Harvard University]

any discussions. In addition, he could stand being in the same room with Thomson. Coffin moved to cut all of Edison's non-power related projects. By today's legal view, the Edison General Electric and Thomson-Houston merger would probably have been knocked down. It gave the new company 80% of the market in lighting and power generation and a true monopoly in streetcars and electric trains. Westinghouse still had one trump card over the new General Electric; GE was legally blocked on AC power generation via Westinghouse's ownership of the Tesla patents.

Chapter 9. Westinghouse and the Tesla System

> I have just seen the drawings and descriptions of an electrical machine lately patented by a Mr. Tesla, and sold to Westinghouse Company, which will revolutionize the whole electric business of the world. It is the most valuable patent since the telephone.
>
> Mark Twain, from a November, 1888 entry

The first three years of the 1890s would be to a large degree define Westinghouse's legacy and well as Westinghouse Electric's destiny. The decade would start with the deaths of his older brothers, Jay and John, and his father's death while staying at Solitude. With these deaths Westinghouse turned ever more into his work. In 1890 Westinghouse had rebuffed the House of Morgan in refusing to join a trust, only to see Edison General Electric merger with Thomson-Houston to become General Electric. At the time the new company of General Electric dominated the market with its DC power system, however, Westinghouse was gaining in the most rural and less densely populated areas. He was having success in Europe, and had sent his young Lewis Stillwell to England to help start British Westinghouse Electric. Westinghouse lost the patent battles for the incandescent lamps in 1891. Still, thanks to Tesla, Westinghouse had General Electric blocked from entering the AC power generation. Finally, Thomas Edison's publicity campaign was putting fear in customers around the country over AC power. The period of 1890 to 1893 saw huge amounts of cash flowing in from all of Westinghouse's companies, but the out flow of cash for AC power development was greater. The electrical power business represented a very different type of cash flow management, which required huge upfront layouts on new projects. For the first time Westinghouse was facing a financial crisis. Throughout the 1890s, Westinghouse failed to manage the balance between debt and development at Westinghouse Electric, which ultimately would cost him Westinghouse Electric.

The financial crisis was not a personal one and would be limited to Westinghouse Electric. Westinghouse Machine Company was shipping steam engines all over the world and even Edison Electric was buying them for DC power generation. His company of Westinghouse, Church, Kerr and Company was experiencing a boom in power plant and factory construction, and was moving into the new field of refrigeration. The nation had also been taken by a safety movement in the railroad, which would make the decade of the 1890s highly profitable for Union Switch and Signal and Westinghouse Air Brake. Westinghouse had excelled in delegation and decentralization of his companies and even Westinghouse Electric's departments, but he remained in control of the finances at Westinghouse Electric to assure his passions and interests could be satisfied. At Air Brake, he had a tough board of directors, which often restricted Westinghouse's money on new projects. Union Switch and Signal Company and Westinghouse Machine were formed because the Air Brake Company did not want to dilute their focus, capital, or efforts in these projects. Westinghouse Electric. From 1886 to 1907, Westinghouse Electric would be the source of funding for new projects. Still, in 1886 Westinghouse Air Brake was the source of Westinghouse's wealth, and he remained active in railroad interests.

One of Westinghouse's biggest supporters in Washington of his railroad projects would be William McKinley then head of the Ways and Means Committee of Congress. McKinley came to Solitude in early 1890 to confer with Westinghouse on railroad safety issues. They would share a passion for industrial safety throughout their careers. Westinghouse would give him a ride on his Pullman car to McKinley's hometown of Canton, Ohio as Westinghouse and his partner Thomas Kerr went on to Chicago. They would immediately become friends, as Westinghouse discussed his new community he was developing at Wilmerding to support his Air Brake operation, McKinley had been a president of the Y.M.C.A. in Canton and noted its value in communities. Maybe more importantly, they shared government support of business while rejecting trusts and a strong belief in "lunch pail" republicanism. Another bond shared was their pride in being veterans of the Grand Army of the Republic. Both men would point to their military service as the highlight of their careers. It would be a friendship that would grow over the years. Basically, a Republican Westinghouse had supported Democrat Grover Cleveland to the surprise of friends. Westinghouse was never active in politics, but he was a Republican going back to his abolitionist roots and became a strong personal supporter of President William McKinley. McKinley, a devout Presbyterian, shared many of Westinghouse's core beliefs. He was even a presidential elector for the McKinley election, and fundraiser. As President McKinley was a hose quest at all of Westinghouse's mansions.

During the McKinley administration he has a personal advisor to the President and purchased a third mansion in Washington D. C., known as Blaine Mansion (Dupont Circle). Marguerite did for a short time became involved in the social and political scene in Washington. Westinghouse's Washington ties became critical to his businesses. Westinghouse's success in his many other industries brought in much needed personal cash, which he used to invest in Westinghouse Electric during the Panic of 1893.

As early as 1890, the financial strain was starting to show Westinghouse Electric. Herman Westinghouse had started to warn his brother of the cash problems relating to the development of the Tesla motor. Herman Westinghouse had become the chief operating officer of Westinghouse Electric, and he saw the hidden problem. On paper in 1890 profits were still strong as well as sales, there was a cash shortage. The books of 1890 showed the company's current assets at $2.5 million and short-term liabilities at $3 million creating a debit of $500,000. A relatively small amount considering the future value of Tesla's patents alone, but George Westinghouse was on a spending binge buying electric patents, bringing the best scientists to Pittsburgh, and hiring young electrical engineers. Westinghouse had developed a strategy of buying patents for precautionary reasons. The strategy frustrated and limited Edison Electric, but was extremely costly. Research and development expenses were accelerating and even their very growth was causing demands for more cash. With the public outrage of the electric chair and the cash problems, Westinghouse put the AC motor research on hold for a while at the advice of Herman, but George couldn't sit on the sidelines long. In fact, in hindsight it would have been just as unwise to curtail research and development at this time. George had three major projects about to launch —the Niagara Falls generating plant, the AC to DC rotary converters and a World Fair contract, all of which would require huge sums of capital.

The rotary converter work was well under way and needed to be extended for the future success of the Niagara project and the use of alternating current. The rotary converter could take AC current and convert it to DC current. This would allow Westinghouse to generate AC power and distribute via high voltage then step it down and convert to DC for use in electric motors, streetcars, and even lighting. This would revolutionize the power generation industry. In the late 1880s there were a number of patents on rotary converters, but Westinghouse's research effort were making improvements and new advances. Westinghouse's newly promoted chief engineer Benjamin Lamme was making real progress in this area. The rotary converter would give the bidding advantage to Westinghouse in major generation projects such as Niagara Falls and the upcoming World Fair, but more research money was needed. The rotary converter, however, had the potential to (and would) knock out DC power generation competition. This represents a rare case where Westinghouse didn't see the full potential of an electrical device. Fortunately, Westinghouse's organization was not dependent on his support to pursue good ideas. Westinghouse knew he had to move forward on the development of the rotary converter as well as Tesla's AC motor. Benjamin Lamme summarized Westinghouse's response:

> Strangely enough, it appeared to me that Westinghouse never took any strong interest in the rotary converter as affording a means for extending the field of direct current traction, and yet this has been possibly the greatest single step in overcoming the early limitations of the 600-volt system. He did not ask me many questions regarding rotary converter development and operation, as he did with other developments. He seemed pleased with the success and its rapid growth after it had passed through its earlier experimental stages.

In the spring of 1890 Herman and George called the board of directors together to address the problem of cash. Westinghouse felt the cash had to be raised; they could not curtail R&D funds in the middle of this technical battle. Westinghouse initial response was predictable from his past financing practices. He suggested a stock issue to raise capital, and he would infuse personal money from his other business profits. The board with the help of Herman Westinghouse, then president of Westinghouse Machine, proposed a reorganization and tile change to Westinghouse Electric and Manufacturing Company, which doubled capital stock. Old shareholders could come in at a discounted price and new issues would raise capital. The general economy had taken a downturn anticipating the Panic of 1893. In 1890 the major bank of Baring Brothers in London failed causing a credit shortage. The strain was even causing Morgan problems to raise money for his upcoming Thomson-Houston and Edison Electric. The offering failed to supply the necessary capital to end the internal crisis. Furthermore, Westinghouse didn't have enough personal cash to pull the company out. Westinghouse was now forced to go directly to the banks for a loan.

A meeting of local Pittsburgh bankers was set for later in 1890. The amount of financing required called for a syndicate of banks, since no single Pittsburgh bank could handle the amount needed. The Andrew Mellon controlled most of the Pittsburgh banking, and with Henry Clay Frick, a deal could be made locally. Mellon and Frick had recently partnered to form Pittsburgh National Bank. Westinghouse approached Mellon in an almost demanding attitude, which Mellon felt was belittling. Westinghouse had tried to avoid allowing bankers in his companies all his life believing that debt to banking institutions is turning over control to outsiders. His New England father had always kept his company free of such debt. Now Westinghouse was torn by his love of technical research and a new view of business practices. Westinghouse was confident that the local bankers would be happy to help because of his enterprises part in the overall Pittsburgh economy. Surprisingly, a number of bankers did not show, which both disappointed and hurt Westinghouse. This appeared to be related to some behind the scenes effort by the House of Morgan. Morgan had started an informal campaign to suggest the weakness of Westinghouse as a businessman. Westinghouse stock was dropping rapidly with all the negativity, which further restricted Westinghouse's ability to raise capital. Morgan was getting almost daily reports by his agents on the crisis of Westinghouse Electric. Even so, Westinghouse had gotten enough of Pittsburgh's bankers together to raise the money. Mellon was still hesitation, but Frick seemed willing.

The dark clouds of the future Panic of 1893 had started early in the 1890s. Bankers such as J. P. Morgan had been expecting the banking crisis. The gold reserves of the Treasury were approaching the legal limit in 1891. The conservative banks of Pittsburgh had been pulling back expecting the crisis. Eventually 575 banks would fail nationally between 1892 and 1893. Pittsburgh banks had been hit hard in the Panic of 1873, and had been cautious. Even Carnegie was having problems raising problem to expand his steel works. Money for civic projects, such as the Allegheny Observatory, was being held up as well. Westinghouse needed money at one of the worst times for American bankers. It was just as difficult for Westinghouse whose Puritan business practices called for debt free operations. He was torn by his desire to advance technology and that of his Puritan roots. After all, his brother Jay was still running his father's company on a debt free basis. Even George had Air Brake and his other companies on a debt free operation, but Westinghouse Electric was in the high tech industry of its day.

The bankers formed a commission to study what would be needed to bring Westinghouse Electric out of the crisis. Westinghouse offered his house Solitude as collateral as well as other personal assets. The physical assets and patent rights of Westinghouse Electric were extensive, which made the loan feasible. The problem became one of control. Local bankers had suggested the problem had arisen from Westinghouse's lavish spending on AC current research. Still, the report was highly favorable suggesting that Westinghouse's assets could easily cover the loan request. One comment of the report told the story, "Mr. Westinghouse wastes so much on experimentation, and pays so liberally for whatever he wishes in the way of service and patent rights that we are taking a pretty large risk if we give him a free hand with the fund he has asked us to raise. We ought at least to what he is doing with our money."[47] They wanted a representative on the board of directors and a general manager to watch the spending. Westinghouse would not agree to this in any form. First he realized the battle he was locked in would require freedom to invest in research. To get the loan and lose the war for AC power generation would end Westinghouse Electric. To all of Westinghouse's organization the task at hand would require Westinghouse leadership unchallenged by banking concerns. Just as important, he was not going to share control, knowledge, or be second-guessed. Andrew Mellon, the Dean of Pittsburgh, wanted some say and control. All the great companies of Pittsburgh had Mellon representatives on the boards, and after all, Mellon was represented on the board of Carnegie Steel. In one meeting reminiscent of his boyhood, Westinghouse stormed out. Meetings continued but Westinghouse now looked to New York as a possible solution.

47 Leupp, 159

Westinghouse was now under pressure from the *New York Times*, which continued to talk of Westinghouse's financial problems. Clearly, Morgan was again using his power to press Westinghouse. The press articles implied poor management, and directed the concern at the financial stocks to assure he could get no relief through the selling of stock. The usually supportive Pittsburgh press and technical journals started to turn on Westinghouse. The claims of poor management were coming out of Morgan's library and they were hurting the stock price and bond ratings. Westinghouse exhibited a great deal of courage and confidence in this crisis. He was confident in solid sales numbers and increasing market share. His strength was in God, family, and friends and it had carried him before.

Probably the most encouraging thing was the offers of Westinghouse managers and employees to help out in the crisis. One example was an offer by T. Gillespie, who had drilled Westinghouse's first gas well, was owned some late bills, but offered to delay them if it would work. A delegation of Westinghouse workers even offered to work at half pay until the crisis was over. Westinghouse had many of these offers, but true to his form he refused and paid all bills in full. This was typical of George Westinghouse, others came first in money matters that went straight back to his Puritan New England training. Westinghouse had begun to feel out New York bankers for a syndicate when he received a call from Charles Coffin, President of the new General Electric. Morgan had the advantage now and wanted to see if Westinghouse would fold. Early on Coffin confided some experience that gave Westinghouse caution. Clarence Barron, famous Business reporter, recalled Westinghouse discussing it with him:

> He [Coffin] told me how he ran his stock down [when at Thomson Company] and deprived both Thomson and Houston of the benefits of an increased stock issue. He was enabled by the decline in stock which he had forced, to make a new contract with both Thomson and Houston, by which they waived their rights to take a new stock in proportion to their holdings under their agreement with the company. [Westinghouse then said to Coffin] You tell me how you treated Thomson and Houston, why should I trust you after what you tell me?[48]

Coffin had told Westinghouse the story to scare Westinghouse, but he underestimated him. Westinghouse was now, however, under real pressure from Morgan and General Electric. General Electric was big enough to control prices and was winning the patent battles. Edison's fear campaign on AC current had been very successful, and Morgan behind the scenes had a publicity campaign touting Westinghouse's mismanagement driving Westinghouse stock down. Westinghouse still would not sell out his company. Rumors, probably originating from Coffin continued to circulate around Wall Street. Attacks on Westinghouse started to surface in the financial papers, "From all the stock-market sub-cellars and rat-holes of State, Board, and Wall streets crept those wriggling, slimy snakes of bastard rumors . .

48 Clarence Barron, *More They told Barron, Notes of the late Clarence Barron* (New York: Harper & Brothers, 1930)

. George Westinghouse has mismanaged his companies . . . is beyond extrication unless by consolidation with General Electric."[49] This was exactly the technique that Coffin and Morgan used to drive down the stock price of Thomson-Houston that was used to merger it into General Electric. Westinghouse did go down, but Westinghouse weathered it.

Short of a merger then Coffin offered him an arrangement to fix prices. He asked Westinghouse: to raise with him the price of lighting from $6 per lamp to $8 per lamp. Westinghouse said that $6 gave fair manufacturing profit, but Coffin said that the Thomson-Houston policy was boodle and that it cost in payments to officials about $2 per light, and if they made the price $8 they could spend $2 with the aldermen, etc., and still get their manufacturing profit.

But Westinghouse wanted nothing to do with it. Meanwhile, he was making trips to New York to obtain loans of his own.

Westinghouse remained hopeful as well as resistant to any plan that would give control to outsiders. The headlines of the Pittsburgh Press on November 16, 1890 certainly would of shook his resolve. The paper announced the failure of England and the world's greatest bank Baring and Brothers. Actually, Westinghouse had just moved into his new Erskine mansion with Marguerite and George III. He received by telegram and immediately readied his Pullman car for a return to Pittsburgh. The fall of such a great bank would certainly tighten money further even in remote banking centers such as Pittsburgh. For J. P. Morgan such a crisis offered opportunity. The Baring Brothers had been a major international rival of Morgan. The Union Pacific Railroad had been borrowing from the Barings to stay out of clutches of Morgan. The Union Pacific collapsed selling out to Jay Gould, an ally of Morgan and pulled down all the railroad stocks. By early December, Morgan had patched together a gigantic railroad trust, diverting his attention from Westinghouse. Morgan proudly announced in December 1890 I am thoroughly satisfied with the results accomplished. The public has not appreciated the magnitude of the work. Think of it —all the competitive traffic of the roads west of Chicago and St. Louis placed in the control of thirty men."[50] Morgan's excitement of trusts was not shared by men such as Westinghouse who saw competition as the engine of capitalism. Still, for Westinghouse the timing of the Baring Brothers probably saved him from a Morgan effort to take over his company.

By mid-December of 1890, Westinghouse stock had dropped from $50 dollars a share to $13 dollars, basically limiting Westinghouse's only potential source of capital to bank loans. A number of small lawsuits to get bills paid arose to further strain Westinghouse. In early January of 1891, the New York Times was suggesting bankruptcy and receivership. Morgan had Westinghouse in a corner, again. Morgan, however, was not at full strength having just completed the General Electric merger and the huge railroad trust. The

49 Margaret Cheney, *Tesla: Man Out of Time* (Barnes & Noble Books, 1993), 47
50 Strouse, 306

bringing together of giants during the period 1890 to 1892, had strained the House of Morgan. Morgan realized that Westinghouse was not going to join a trust of any kind, and with diversified Westinghouse debt obligations he had no way to take over in the case of a Westinghouse failure. In addition, Morgan was looking to reorganization of his railroad trust with the unsettled financial times running up to the Panic of 1893. Morgan had been anticipating this coming crisis as an opportunity to use debt to take over more railroads. Morgan's diversions allowed a window of opportunity for Westinghouse in the New York markets. Westinghouse during the period had worked with the banking house of August Belmont to put together a syndicate to pull Westinghouse out of the crisis. August Belmont Jr. was one of the few bankers powerful enough to act without Morgan. August Belmont the elder, who had died in 1890, was a personal enemy of Morgan. The Belmonts were Democrats as well, opposing the Republican dominated Morgan network. Belmont Jr. had better relations with Morgan, but he was still willing to put together a syndicate for Westinghouse. The firm of August Belmont also had solid European ties to the Rothschilds. The Rothschilds had dealt with Westinghouse in the air brake business in the 1870s and was looking to expand Westinghouse's electrical business in France. Belmont had Morgan's skills and looked beyond mere loans, wanting to form a stronger Westinghouse Electric capable of competing with General Electric. Belmont was also in agreement with the winning of two projects under bid — Niagara power generation and the 1893 World's Fair. Westinghouse needed more than the half million to resolve the immediate problem.

When Westinghouse went to New York on a bitter cold January day, it would be his first meeting with the big city bankers. August Belmont Jr., was extremely cordial and helpful to Westinghouse. August Belmont was a pure moneyman lacking the colossal trust dreams of bankers, like Morgan. The deal of Belmont would bring in two companies — United States Electric Lighting Company and Consolidated Electric Company. Both of these companies had lost a major court battle with Edison General Electric over the incandescent lamp filament. Westinghouse through leasing arrangements had control over United States Electric and Consolidated, but had not purchased the assets. Westinghouse had been purchasing shares of United States Electric Lighting since 1888, but lacked control. United States Electric was one of Edison oldest competitors. It was founded the same year as Edison Electric (1878) by prolific inventor Hiram Maxim. Even Morgan had been looking at United States Electric because Wall Street believed all it needed was more technical expertise with Hiram Maxim having left the company. The Belmont syndicate brought United States Electric and Consolidated into Westinghouse Electric and Manufacturing Company. United States Lighting was a major competitor in both arc and incandescent lighting with Westinghouse, Edison, and Thomson-Houston. The stock issuance required some sacrifice from the Westinghouse stockholders to reduce liability. Westinghouse pulled together a small army of friends and employees to buy preferred stock and close the deal. The end result was that debt was reduced and interest payments. Belmont asked for no over riding control in the company as the Pittsburgh bankers had. The Belmont deal created an enemy in Pittsburgh banker, Andrew Mellon, which would hurt Westinghouse in the Panic of 1907. It was a triumph for Westinghouse, and freed him to pursue Niagara and the World's Fair as he saw fit. In hindsight, Westinghouse might have been helped by some financial review in Westinghouse Electric.

Niagara Falls for years was being called "white coal" as the energy of the falls was enormous. The Niagara power project really goes back to 1890 with the formation of the Cataract Construction prior to the formation of General Electric and Westinghouse Electric and Manufacturing Company. Edison was one of the early investors in the Cataract Construction, because of the problems in lighting New York City. Edison's DC system would require thirty-six power stations to service all of New York City. He was far from that goal of 36, as citizens were complaining about the noise and smoke of his steam driven central power stations. Vibrations were being felt in near by buildings. Threats of lawsuits were rising, and even available land for these stations was in short supply. Edison, of course, was hoping for a DC power plant at Niagara. In 1889 Edison had suggested a plan to transmit DC power from Niagara to New York, but many engineers questioned the feasibility. The president of Cataract was Edward Adams, a director of Edison Electric and a Morgan partner, appeared to give Edison the initial advantage. Considered a physically unimpressive, he was an attorney and banker of significant stature. Edward Adams was Morgan's hand picked man serving as a Morgan representative on a number of railroad boards. Adams was a Boston banker and blue blood being a descendant of Presidents John and Quincy Adams. He had been a key investment banker in Morgan's formation of the railroad trust. As an attorney, he had worked with Coleman Sellers engineering firm in Philadelphia. Adams was forced to sell his shares in Edison Electric to be impartial. He also established an International Niagara Commission in London because of Europe's experience with hydroelectric power generation. Adams put the famous scientist Lord Kelvin in charge of the Commission, which he hoped would favor DC. Adams remained in contact almost daily with J. P. Morgan while on the project.

The commission's first reports favored DC probably due to the political pressure applied and Lord Kelvin's dislike of AC current. The Commission then suggested a contest for the best plans for the harnessing of Niagara Falls. At first Westinghouse's man on the committee in Europe, Lewis Stillwell agreed to it. Westinghouse, however, refused the proposal feeling he would need to reveal his technical advantages for the small sum of $3000. Westinghouse suggested he would be selling $300,000 of technology for $3000. In 1891 two major demonstrations of AC power came on line both related to Westinghouse. First was an alternating current transmission from Lauffen to Frankfurt, a distance of over 108 miles. In Portland, Oregon, Westinghouse had built the first American hydroelectric AC power plant. Willamette Falls generated the power at 3,300 volts then moving the current 13 miles to Portland stepping it down to 1000 volts, and to customers stepping it down to 100 volts. Another plant of Westinghouse design used a waterfall in Tivoli to produce AC current and sent it twelve miles to Rome. In addition, the generation was polyphase, which made the Tesla motor operational, but even without an AC motor; Westinghouse was producing rotary converters to turn AC to DC for motor applications. Westinghouse had a commercial advantage, but Lord Kelvin still supported DC. Adams and Morgan realized AC was starting to win.[51] The battle now turned to the World Fair contract that would probably resolve the Niagara contract as well.

In 1890 requests were made for bids to light the upcoming World fair in 1893. Westinghouse saw it as a key part of his strategy to win not only the Niagara contact but the war of the currents in general. He had two problems in such a strategy. The patent battle for the incandescent lamp was still in court, but Edison seemed favored. Westinghouse was also up against the House of Morgan with its open control of General Electric. Westinghouse felt he needed the World Fair contract to overcome the publicity campaign against AC current. Had the fair been in New York, Westinghouse would had lost, but this was Chicago. Chicago had independence in banking and in the press. Westinghouse also still had some technical advances to make, but he had some of his best engineers on the project in Pittsburgh. Tesla was in Pittsburgh at the time under contract, and he had his young engineer Benjamin Lamme on the project too. Organization was becoming the strength of Westinghouse Electric. While Edison was still making points in the public press, the technical journals had started to favor AC. Westinghouse engineers dominated the technical societies as well. The Edison organization was showing its weaknesses, but General Electric was moving quickly to a Westinghouse-style research organization.

51 Charles Scott, "Nikola Tesla's Achievements in the Electrical Art," *AIEE Transactions*, 1943

Chapter 10. Westinghouse Goes to the Fair

Westinghouse seized the opportunity to use the upcoming World Fair to advance AC current as well as his company. He reasoned the fair would highlight his system to the world. The Chicago World Fair was to be called the Columbian Exposition in honor of the 400th anniversary of Christopher Columbus' landing in America in 1492 (to be held a year late due to the presidential election). The cities of New York, Washington, D.C., St. Louis, and Chicago were all in contention. Chicago won out on April 25, 1890. The competition caused a New York editor to call Chicago "that windy city." President Harrison signed the act designating 630 acres in Jackson Park and the Midway for the World's Fair. Westinghouse was excited about the prospects, having enjoyed himself at the Philadelphia Fair of 1876. The 1876 fair had been a major marketing success for his air brake, and he believed this fair could do the same for AC current. Edison had also been triumphant at the Paris Exposition of 1889. The Paris Exposition had used Edison's system to power 1150 arc lamps and 10,000 incandescent lamps. The bid specifications for Chicago called for 5,000 arc lamps and over 90,000 incandescent lights. The power requirements were ten times that of Paris. The technical requirements would be challenging to both General Electric and Westinghouse Electric. It was assumed that the newly formed General Electric would win because of the financial demands. Many in the Westinghouse organization were opposed to the costs. Herman Westinghouse argued against it, realizing that even with loans this could bankrupt the company. George was determined to make it work, against the advice of many. Marguerite seems to have been an important supporter of this project, even offering to make any personal financial sacrifices needed. Bids opened in April 1892, after Westinghouse had put together his financial syndicate with Belmont, but Westinghouse was facing an even bigger technical challenge.

On the fear campaign of Edison, Westinghouse gained some relief, as Morgan forced Edison out of General Electric. Morgan and Coffin at General Electric now realized that the future could include AC, and they wanted to hedge their bets. Still, the campaign continued during the bid period. Westinghouse, however, was still in a court battle over incandescent light bulbs. A Westinghouse loss would leave him with no incandescent lights for his system. Westinghouse had sent a team of engineers to Chicago to review and report. The cost would be enormous, but technically Westinghouse engineers were confident. The bid was arranged in several engineering phases. First, bids were requested for the arc lights. Westinghouse was not fully invested in the arc lighting systems and chose not to bid. The newly formed General Electric had control and was the lone bidder capable of the project. Coffin feeling the power of Morgan's electrical trust, put it to the fair committee with a bid of $38.50! Recent fairs had installed arc lights at $11 to $13. Coffin had made a major blunder in this strategy. Morgan did not control the Chicago press, and they reacted to the gouging with vengeance. The Chicago press had already been on the forefront of a national anti-trust movement, which Coffin played into the hands of. This was not Morgan territory, and Coffin's monopolistic behavior was seen as very favorable. The committee was also outraged and worked out a very profitable deal with a group of smaller companies for $20 a lamp.

Sealed bids for the bigger power system were due in April 1892 just prior to the formal formation of General Electric on April 15, 1893. All were controlled and related General Electric's bid on the Fair except one. Westinghouse's bid was initially as a subcontractor to South Side Machine and Metal Works of Chicago. It is not clear whether Westinghouse was initially working with South Side or brought in afterward. Clearly, South Side realized they would have to subcontract to Westinghouse. Westinghouse may have also used it as a feint, drawing Coffin into a higher bid, which appears to have been the case. In any case, General Electric took the bait and used the monopoly playbook. The General Electric bids ranged from $13.98 to $18.51 per lamp using Edison's DC system. The South Side bid was an amazing $5.49 using Westinghouse's AC system. The selection committee called in both Westinghouse and General Electric to go over the details. Some of the committee seemed to be strongly in the Morgan and General Electric camp, but the Chicago press favored Westinghouse. Westinghouse earlier had been forced to hire a public relations man to battle the Edison DC campaign. Ernest Heinrichs was that man, and he was well suited for the job having been an industrial reporter with Pittsburgh's *Commercial-Gazette* and *Pittsburgh Dispatch*. Heinrichs had good connections in the newspaper business, and excellent writing skills. He had only recently came to the United States from Germany, and had solid ties in Europe as well. Heinrichs had probably edited many of the Westinghouse 2-6 page replies to journals heavily. Westinghouse hated to write with the few exceptions were descriptions of his product were required.

The Fair officials asked for new bids, and Westinghouse left for Chicago. Heinrichs set up a press conference at the Auditorium Hotel where he was staying. Personable Westinghouse was the champion the Chicago press was looking for. He won them over, and they applied the pressure. Westinghouse took his earlier low bid to $5.25 per lamp. General Electric bid $6, but on realizing they were well over Westinghouse's bid, Coffin moved to argue technical points further. Some committee members argued that Westinghouse could not deliver if General Electric won the ongoing patent battles, and furthermore, that Westinghouse lacked any bulb manufacturing plants and General Electric could refuse to sell bulbs to him. Westinghouse used a friendly press and his dynamic personality to win over the committee. In addition, he assured the committee he could produce the needed bulbs himself. The public and anti-trust press had forced the committee to go with Westinghouse. Even Westinghouse's closest associates were questioning his wisdom, but Westinghouse knew if he lost the Fair bid, he would probably lose the Niagara project. Morgan seemed to be willing to give Westinghouse the victory at Chicago while re-doubling his pressure on the Niagara project. But Morgan had his hands full, as the initial crisis of 1893 was under way in New York banking circles. The newly formed General Electric had a huge amount of debt that would have to be protected in the emerging crisis.

Westinghouse was excited, but other than his wife, many in his organization did not share his enthusiasm. Westinghouse's Solitude was buzzing with dinner parties, of which George proudly showed his plans for the upcoming fair. It became a nightly routine to retire with his quests to his study and spread out the drawings on his billiard table. Nikola Tesla had become a favorite of Marguerite and George during these challenging times. Tesla planned to have his own exhibit under the Westinghouse banner. Benjamin Lamme and Albert Schmid were also frequent dinner quests reporting on activities and progress. These were very tough times, as Westinghouse probably had won the fair bid at a million dollar loss to the company, and his obligations to New York bankers were growing at a disturbing rate. His brother, Herman, did not share his enthusiasm for the project in light of the debt it was creating; yet the Belmont banking syndicate seemed capable of maintaining the cash flow for the project. Westinghouse's love of engineering seemed to keep him free from the financial concerns. His daily retreats to Solitude, Marguerite, George III, and his dinner quests renewed his spirits. His dinner quests were usually scientists, engineers, and friends. Marguerite steered the conversation from business, which was not difficult, as George enjoyed just being with friends.

During the period of 1889 to 1893, the Westinghouse family was in a state of flux. Both George and Marguerite's families were extremely close and they loved having family at their homes. As always, family and friends often mixed with the rich and famous at their dinner parties. Eighty-one year old Mrs. Westinghouse had moved in with them after the death of her husband at Solitude. His parents often stayed months at a time at Solitude, but Erskine was closer to their roots. The Erskine mansion was evolving into their summer home. Westinghouse often amused himself in the wiring of the Erskine mansion to the frustration of Marguerite. Westinghouse kept the wires exposed, running along walls and ceilings so he could make changes readily. Marguerite saw an improvement in her health with the Berkshire living. Her mood improved as she could wear white dresses, and could actually see the sun routinely. Westinghouse continued the dinner parties regardless of which mansion they were at. We have a further insight into his family life from the family secretary, Walter Uptegraff, who joined the family in 1888. Uptegraff observed the following in his employ at all the Westinghouse's homes:

> After the quests had gone and the family had retired, he would go to his study where he would work on his inventions until one or two o'clock in the morning. During these quiet hours he always liked to have someone with him to whom he could open his inmost self without resistant; to explain what he was trying to accomplish by showing the drawings he was working on; to compliment the listener by inviting suggestions. He required but four or five hours of sleep and would come down to an early breakfast bursting with energy, eager to get at the day's work as quickly as possible.[52]

During the 1880s, when he was at Solitude, he would eat at 6am and take the train to his office on Penn Avenue. He was behind his desk at 8 am. When his plants moved in the 1890s to East Pittsburgh and Wilmerding, he liked to go to the plant first and then take the train back to his office in Pittsburgh.

His routine changed in the spring of 1892 with the winning of the major portion of the lighting contract at Chicago Fair. He was locked in a patent battle with General Electric, but worse yet was that the General Electric trust would not sell him any lamp globes. They now hoped to embarrass Westinghouse by failing to deliver at Chicago. Thomas Kerr, his patent lawyer, was sure that Edison would win the lawsuits and end up controlling all of the incandescent lamp patents in the world with the exception of the old Sawyer-Man patent. Westinghouse had to decide to move forward in manufacturing an Edison globe facing the loss of the lawsuit or try to develop a new one. In any case time was short for he needed to set up his own glass factory since the trust controlled all globe capacity in the U.S. Westinghouse called a counsel at Solitude to brainstorm the problem. The meetings went on for days and included his best technical and legal minds. Thomas Kerr brought in another expert in electrical patents, Leonard E. Curtis. On the technical side he had Schmid, Lamme, Tesla, and his incandescent department head Frank Stuart Smith. Albert Schmid had been one of the few supporters of Westinghouse's decision, but he assured Westinghouse he would

52 A. G. Uptegraff, "The Home Life of George Westinghouse," July 1936, page 2, George Westinghouse Museum Archives

pull his research team into the effort. Tesla, Westinghouse's other ally was in Europe, but they were in contract daily by telegraph. Westinghouse did own the old Sawyer-Man lamp patent of 1888. It was a crude device that lacked reliability in operation. Kerr felt that an improved Sawyer-Man lamp would fall outside the Edison patents. Westinghouse had a research department dedicated to incandescent lighting and had developed a modified two-piece lamp based on the Sawyer-Man patent. Unlike the Edison lamps that consisted of a vacuum-sealed one-piece bulb, the two-piece consisted of a bulb with a stopper made to fit the neck. The carbonized filament needed protection from air, which caused it to burn out; this protection could come from a vacuum or an inert atmosphere. Besides the two-piece lamp, the Westinghouse lamp used inert nitrogen gas in place of Edison's vacuum. Thomas Kerr, Westinghouse's patent lawyer, was confident this was a unique approach that at worst infringed on the Sawyer patent, which Westinghouse owned. The manufacture of this stopper bulb would require extremely close grinding tolerances to assure a tight fit to prevent the leakage of nitrogen. A tight seal was needed to assure the life of the filament without which a bulb would burn out. Still, this remained Westinghouse's best strategy.

In May he started his own glass factory on Pittsburgh's North Side (then the city of Allegheny). Initially, the glassworks was producing Edison type bulbs while the patent issue awaited action by the Supreme Court. The glassworks was housed in the portion of Westinghouse Air Brake that had moved its operation to a new plant in Wilmerding. He was able to draw on the glassmaking expertise in the Pittsburgh, which was the oldest in the nation. He built a team of expert glass blowers. Westinghouse used the best technology and was even pioneering machine blowing allowing him to produce Edison bulbs at a lower cost than General Electric suppliers, Corning Glass and Libbey Glass. Westinghouse's technology of automated glass making was at the time several years ahead of inventor Michael Owens, which is typical of Westinghouse's application to new ideas. He established the glassworks as a separate company known as Westinghouse Glass. When the Supreme Court ruled in General Electric's favor in December of 1892, Westinghouse was ready to switch to stopper lamp production. The grinding of the glass neck was critical to assure a good fit, and Westinghouse spent a great deal of time at the plant involved in this part of the process. Westinghouse seemed to love the challenge of bringing this stopper lamp on stream to trump the latest move of General Electric. Westinghouse's Fair manager remembered his enthusiasm:

> Westinghouse ... having overcome a great obstacle... was bubbling over like a boy. He explained the operation of the grinders and I saw that the men ... seemed imbued with the idea that it was a game to beat an opponent who held all the aces, and that they were having a lot of fun doing it. He had a sort of magnetic influence on the workmen.... It certainly was a great delight to realize that, in spite of what seemed a hopeless situation,

> "the boss" was going to furnish lamps without paying tribute. He certainly lifted the worry from me.[53]

This was Westinghouse at his best, the enthusiastic engineer with a goal. He had loved the great brake trials of Burlington and London, the decade before. It wasn't Westinghouse alone, however, it was an energized organization. He even had named an overall fair project manager E. Keller. While Westinghouse tackled the grinding issues, his incandescent lighting department continued to experiment with design.

The task ahead of him was enormous, but as in his own case, he thought the organization worked best under pressure. Benjamin Lamme would later reminisce, "He believed, apparently, that most engineers did their best work when hardest pressed; and no doubt, to a certain extent, he was right." He had to bring a glassworks on line and produce 172,000 lamps plus a stream of replacement bulbs for a total of 250,000. Westinghouse was well aware that his stopper lamp was but a quick fix. It did not have the life or reliability of the General Electric lamp. In fact, he had an army of men replacing the short-lived lamps at the world's fair. The stopper lamps leaked nitrogen slowly and that would cause burnout. The price of the stopper lamp met that of Edison's lamp at around 28 cents per lamp. Schmid and Smith were even experimenting with platinum filaments, which could resist burning better, but the cost was prohibitive. The strength of the Westinghouse organization was demonstrated in his effort to prepare for the World's Fair. While Westinghouse, Smith, and Schmid led a battery of engineers on the stopper lamps, Charles Scott, Benjamin Lamme, Nikola Tesla, and W. Rugg worked on revolutionary polyphase AC generators, transformers, and motors. Edison's autocratic research organization could never have handled such a short-term developmental program. General Electric with Edison gone would model their future research and development efforts after the Westinghouse operation, which performed so admirably during 1892-93.

The spring and early summer of 1892 provided some relief from the negative publicity campaign by General Electric. With the merger and reorganization of General Electric, Edison was forced out of an active role. Edison was not favorably disposed to anyone at General Electric. The newspapers almost hailed his ouster, pointing to his management shortcomings. This campaign was typical of Morgan takeovers, which tried to present his takeovers as necessary to improve management and stockholder equity. In this case, it was a failure, and sympathy for Edison grew. From his earliest days, Edison had courted the public. General Electric was facing a public relations problem and an upset Edison going into the fair. Edison was threatening to not participate in the World Fair. General Electric needed Edison's goodwill to be successful at the Fair. Since Edison would not talk to Coffin or Morgan, they chose Fredrick Fish to act as mediator. Fish spoke directly to Edison, "Tell me just how to meet your views, and we will meet them." The price was

53 E. E. Keller, "George Westinghouse Memories," April, 1936, Volume 3, Box 1, Folder 8, George Westinghouse Museum Archives

high. Edison would receive 12.5% of all profits from lamp sales, and a salary of $600 a week to support his laboratory work. General Electric brought Edison back into the fold, provided he would not carry on the technical assault of Westinghouse. It is interesting that during this period a famous engineer and admirer of Edison joined one of Edison's companies. Henry Ford became an engineer at the Detroit Edison Illuminating Company, and by 1896, he would be chief engineer.

By the summer of 1892, Westinghouse had his glassworks under way and his organization arranged in order to meet the challenge. In Pittsburgh and for awhile the nation's businesses focused on the labor unrest at Homestead visible from Westinghouse's Penn Avenue office. The newly formed Amalgamated Association of Iron and Steel Workers was ready to challenge Carnegie's huge Homestead Works. Andrew Carnegie was in Scotland and Westinghouse's neighbor and poker club member Henry Clay Frick was in charge. On July 6, 1892, the differences broke out into violence between the workers and the company hired Pinkerton guards. The gunfight resulted in the deaths of seven workers and four Pinkerton guards. Some of the deaths were caused by an angry mob of workers and family members making the guards walk a gauntlet of beatings. Two weeks later, Henry Clay Frick was knifed in his office by an anarchist. A few days later, while recovering at his Homewood home, Frick's new son (born on the day of the Homestead gunfight) died. Westinghouse and Marguerite were there to help in a time when few in Pittsburgh cared whether he lived or died.

For months, labor and gang unrest increased around the Pittsburgh area. The main manufacturing operations of Westinghouse Electric as well as the glassworks in Pittsburgh's North Side (Allegheny City in 1892) were in the "tough district." Patrol wagons were common, but Westinghouse workers remained dedicated to the company. Westinghouse workers were well cared for in stark contrast to that of Carnegie's mill workers. In 1892 a mill worker worked a ten to twelve hour day, seven days a week. Christmas was the only day scheduled off with the exception of business downturns and January maintenance days. Westinghouse employees were better paid and had Sundays off, although this summer the plant and engineers were voluntarily working on Sundays to meet the timetable of the Fair. Both Lamme and Rugg reported working from early Sunday morning to late at night during the spring and summer of 1892 breaking only to watch when the patrol wagons went by.

Tesla arrived back from Europe in September and Westinghouse and Schmid met him at the train station. They had much to go over with him. They convinced Tesla to commute from New York to Pittsburgh through the end of 1892. Westinghouse was committed to generating polyphase AC for the Fair, and Tesla's 60 cycle transformers and dynamos would be needed. The polyphase AC would be the cornerstone of his Niagara bid. Westinghouse realized that Tesla was critical to their development even though many in the organization resented Tesla. In addition, Schmid, Scott, and

Lamme went to New York often to discuss designs with Tesla. Tesla did a terrific job in bringing Westinghouse's future polyphase AC system on line. The Westinghouse exhibit would have a huge banner hailing, "Westinghouse Electric & Manufacturing Co. Tesla Polyphase System." in addition, Tesla was given space for his own exhibit, in which he built one of the first neon lights hailing Westinghouse. The help of Tesla in the latter days of 1892 helped build the confidence of George Westinghouse as the first wave of the Panic of 1893 began. In the midst of major railroads failing and tens of thousands unemployed, Westinghouse borrowed against the Belmont line of credit. Eventually, it was estimated that Westinghouse lost millions on the fair, but it had also launched Westinghouse Electric as the major competitor of General Electric.

December of 1892 brought a new attack by General Electric. On December 15, the courts handed down a decision on the incandescent light, as expected, it was a complete victory for General Electric and Edison. The Westinghouse Sawyer-Man lamp remained outside the decision, so Westinghouse was satisfied that he could meet the fair requirements with the stopper lamp. Westinghouse hoped this would be the end of the "Seven Years War." George and family went to New York to do a little business and shop a few days before Christmas. While there Westinghouse caught wind that General Electric lawyers had left for Pittsburgh. Westinghouse immediately called Kerr and readied his Pullman to return to Pittsburgh. Unable to reach Kerr, he got in touch with George Christy by wire to get over to the United States Circuit Court in Pittsburgh. Christy was able to meet and surprise the General Electric lawyers on December 24 at the courthouse. The General Electric lawyers were ready to file for an injunction to stop the production of the Sawyer-Man lamps by Westinghouse. If they could stop production for even a few weeks it could have caused Westinghouse to be unable to light the fair. Since Westinghouse had the contract, this was a direct attack to hurt and embarrass Westinghouse Electric. Thanks to Westinghouse's tip, his lawyer Christy came prepared to head them off. He first got a postponement until after Christmas. This allowed Westinghouse to prepare the necessary blueprints and technical data. The judge ruled no basis for any infringement until after Christmas. Precious time had been saved, yet General Electric would continue to harass Westinghouse, fortunately to no avail. Westinghouse never could understand the nature of the attacks; he had always been open to a compromise. Morgan appeared to be the force behind the whole strategy. Charles Coffin proved to be a petty man, lacking the compromising skills of Villard.

George, Marguerite, and George III attended the opening day festivities of the Chicago Fair; as was his style, Westinghouse was in the background spending the day with family. It was a day of triumph; only a few weeks before, he had received a telegram that the International Niagara Commission had selected AC current for the fall. No doubt the Fair's use of AC had turned the tide. General Electric would now be on the other end of a patent feud. Westinghouse owned the patents, but he was well aware that General Electric had also developed an AC power system to compete. Even more disturbing was the recent discovery of spying at his Pittsburgh operation. This happened on May 9, two days after the Commission decided on AC current. Initially, Westinghouse thought the leak might have been with the Commission with so many Morgan people around, but a search warrant showed the General Electric had corporate spies involved. A draftsman was fired for selling blueprints to General Electric as well. General Electric argued they were protecting against patent infringements by Westinghouse. Espionage of this type was not uncommon in the Gilded Age and eventually a Pittsburgh jury would split on the decision. The Niagara Commission certainly leaked some through the normal course of comparing and visiting the plants involved. Informal talk at technical conferences was probably another source of leaks; by 1893 the AC system had been a topic of discussion for years. Westinghouse was still comfortable on opening day that he would win the patent battle.

Westinghouse proved right that the Chicago fair would define America; Westinghouse would be remembered always for this great Exposition. The Jackson Park site covered over seven hundred acres, and was enough for over sixty thousand exhibits. One out of every four people in the United States traveled to Chicago to see the fair. Its cost of $25 million would still bring in $2.25 million. A total of twenty-seven million visitors from all over the world passed through the gates. The famous Ferris Wheel, designed by Pittsburgh engineer George Ferris, could carry 2000 people at once. There were 36 trolley size cars with forty stools (and like in a trolley some passengers stood). The ride cost 50 cents and with loading and unloading the two revolutions took 20 minutes. The Ferris Wheel dominated because of its 250 foot steel structure. The Ferris Wheel was the favorite of Westinghouse and little George as well as most of the fair goers, and it was estimated that 1.5 million people rode it. There were over 65,000 exhibitors and many, like Westinghouse, would make their name at the fair, such as Juicy Fruit Gum, Kodak box cameras, typewriters, gasoline engines, Aunt Jemima Syrup, Cracker Jacks, Libbey Glass, Gillette's safety razor, Jell-O, Postum, Cream of Wheat, Kellogg's Shredded Wheat, and Pabst Beer. The fair introduced diet carbonated soda, hamburgers, exotic dancing, and postcards.

The fair would herald the birth of electrical power, featuring arc lighting, incandescent lights, electric trains, electric boats, electric motors, electric cranes, lighted fountains, telephone service, electric fire alarms, and endless futuristic appliances. The fair construction extensively used electric power as described in this article in the September 1983, issue of *Cosmopolitan*:

> Generators ran day and night seven days a week, operating motors in the daytime which furnished power for the saw-mills, hoists, pumps and painting machines, and at night grinding out light, so that the construction could be carried on day and night where necessary, and the engineers and draughtsman could lay out work for other days and nights. Electricity helped to prepare the material, to hoist the heavy beams and trusses, to

paint the buildings, and at the same time to prolong the labors of the overworked engineer and mechanic.

The Edison DC system powered the arc lights and powered a variety of small motors used during the construction. In the earlier days of construction DC motors were used to pump water out of the swamp site selected. These DC motors drove tool sharpeners and sawmills. Westinghouse, observing the extensive use of DC motors in construction decided to highlight his own exhibit with the polyphase AC motors. Westinghouse was often on site to see this wondrous engineering feat unfold, and during the fair itself he could often be seen with George III and Marguerite. Westinghouse because he avoided photographs and the use of his name, had a great deal of anonymity. It would be unlikely that even the average Pittsburgher at the Fair could recognize him. The Westinghouse family stayed several weeks with little George getting special permission to be absent from school. It was said that George Jr. tutored George III every night after returning from the fair. The illumination of the Fair, buildings, and Lake were worth the trip to Chicago, but engineers marveled at much more. The electric train for the fair had six miles of track with fifteen trains of four cars, each able to hold one hundred passengers. The train had a block signal system of Westinghouse's design. Similarly, fair goers could ride on a fleet of fifty electric boats.

The Westinghouse powerhouse was revolutionary and the lighting plant was the largest power station in the world in 1983. The nucleus of this station was six pairs of 500 horsepower generators. Lacking a fully developed polyphase generator, single-phase pairs were used to produce a similar effect. The pairs had wiring in 90 degree out of phase, producing a simulated Tesla polyphase, 60 cycle AC current. This idea of pairs was Westinghouse's innovation, and it was part of his determination to fully develop and apply an AC induction motor. It was the first step toward true polyphase, and represented a longer-range development plan for the company. These pairs weighed 150 tons each. It was the engineering triumph that Tesla had theorized and Westinghouse Electric brought to reality. It was also a triumph of Westinghouse's management and development by set objectives. The generators were driven by an Allis-Chalmers oil fired steam engine, which reduced the smoke. In addition, there were several Westinghouse 1000 horsepower steam engines. Herman Westinghouse had been working on diesel engines to prepare for the fair. Westinghouse transformers stepped down the voltage at points of application. The new rotary converters were used to change the alternating current (AC) to direct current (DC) for use by the electric railway. A number of the Tesla polyphase motors that were being used and were exhibited would be the basis of Westinghouse's future electric motor business. These rotary converters of Benjamin Lamme would eventually change the nature of the competition. The Fair project had motivated development at all of Westinghouse's companies. One project was a natural gas engine to produce electricity by Westinghouse Machine Company. His Philadelphia Company and Westinghouse Machine had jointly developed this engine.

The horsepower of the developmental gas engine could not help him at the fair, but its development became a key part of Westinghouse Machine Company sales to the steel industry. 5000 horsepower gas engines would be used in blast furnaces and rolling mills the world over.

Charles Coffin's General Electric tried to divert attention by highlighting Edison —the wizard of electricity. Edison was a reluctant participant, having lost the company to Coffin, yet Coffin needed the publicity Edison would bring to the newly formed General Electric. Outside Chicago Westinghouse had very little public recognition. General Electric made good use of their small arc lighting contract, illuminating the fountains and many of the buildings. The 2000 horsepower general Electric DC power plant helped power the Intramural Railway that circled the Fair grounds. General Electric powered the electric gondolas on the lake using rechargeable Brush batteries. Their exhibit told the story of Edison's conquest of incandescent lighting with visits from the wizard himself. While Westinghouse could roam the Fairgrounds unnoticed, Edison was like a rock star. The General Electric exhibit featured their new electric locomotive known as the "Titan of Traction," and the historical lighting of the battleship "Illinois." General Electric refused to acknowledge any successes of Westinghouse and the Fair. Some fifty years later in the 1941 company biography, *Men and Volts: The Story of General Electric,* Westinghouse is left out in their analysis General Electric's principal competitors in that year were Stanley Electric Manufacturing Company, the Siemens and Halske Electric, the Brush Electric Company, and the Fort Wayne Electric Company." Even after Morgan wrestled control of Westinghouse Electric from Westinghouse, the cultural differences prevented him from having a united electrical trust.

Charles Coffin would struggle for years because of this setback dealt him by Westinghouse. Morgan never fully believed in Coffin after the setbacks either. At the end of the Fair, Coffin headed up a company in organizational disarray. Coffin had come from Thomson-Houston and Morgan had hoped he could pull the merged company of General Electric together. Coffin had been a concession of Morgan to get Thomson-Houston to sign on with the formation of General Electric, but Coffin was too used to being in charge. Typical, however, of J. P.'s strategy of Morganization, an executive committee dominated by bankers was put in place to pass on Coffin's ideas. Coffin was a dramatic and hard charging manager and Morgan instituted similar committees in sales, engineering, and manufacturing to rule in Coffin. Morgan hoped to allow Coffin to hang himself, allowing General Electric to flounder in the Panic of 1893. Company indebtedness played to Morgan getting more control. General Electric in 1893 had "to pay unheard of rates of interest for money with which to meet its weekly pay roll week after week." Morgan created a crisis in the debt market, which allowed him to leverage more control from both the old Thomson-Houston and Edison Electric shareholders. Coffin was able to remain as President, but his control was greatly diminished. Morganization also required a reorganization to bring central control versus the reneged regional offices. These internal struggles helped Westinghouse as the struggle for the Niagara contract continued.

When Westinghouse returned to business in the fall of 1893, his desk was covered with papers. He was exhausted, but now had to turn to another battlefront. The most pressing was the Niagara project and the building of a bigger factory for Westinghouse Electric. In addition, he had continued to mount debt in the midst of the Panic of 1893. The steel mills of Pittsburgh were short orders and that was bringing down the entire regional economy. The war with General Electric was far from over, and Morgan's dream of an electrical trust would not go away. Westinghouse did take some time to relax and enjoy his newly created community of Wilmerding as well as Erskine Park. In the end estimates of the cost of the fair ranged from a million dollar loss to that of $16,000 reported in the Westinghouse Annual report, adjusting for the sale of the generators after the fair. Sales and profits, however, doubled within a year. Westinghouse Electric and manufacturing had also gained world name recognition. Westinghouse's success allowed him to spend some time on three of his other companies that were experiencing rapid growth —Westinghouse Air Brake, Westinghouse Machine, and Westinghouse, Church, Kerr, and Company. In addition, Westinghouse was involved in the development of a model industrial town at Wilmerding, Pennsylvania.

Chapter 11. Wilmerding, America's Company Town

The War of the Currents had overshadowed the launch of another project for Air Brake in 1890. The company town of Wilmerding would be of Westinghouse's design. Wilmerding would become an international model for industrial towns. In its early years reporters from all over the world flooded it. It was hailed by reformers and socialists as the ideal model for industrial towns. It was despised by the bankers of the time as misguided capitalism. It stood in stark contrast to the many surrounding steel towns of Braddock, Homestead, Duquesne, and McKeesport. Flowers, birds, insects and hardwood trees prospered, while neighboring towns favored only the hearty sumac trees. Houses were well built and surrounded by beautiful lawns, parks, and gardens versus the slag dumps in the neighboring steel towns. Social, educational, and recreational services and opportunities abounded. A truly symbiotic relationship between company and town existed that was unknown in the Victorian industrial world. This relationship reflected the kindness and humility of its founder-George Westinghouse. Yet, Wilmerding was no socialist experiment for Westinghouse, but a pragmatic example of capitalism. Westinghouse put it best, "I believe in competition, it is the essence of a free economy. I think employers should compete in the improving the lot of their workers as well as in making of more and better goods at a cheaper price. It strikes me as common sense that when men are happy and comfortable they produce more and help make a better profit for the company."[54]

Wilmerding did not exist until 1890 except as a rest stop on several intersecting Indian trails. The valley made for a natural transportation course being a node on the Raystown, Catawba, and Warriors path as it is today for routes 30, 40, 22, and the Pennsylvania Turnpike. The beautiful valley has attracted civilization for centuries, the he earliest being the Monongahela people around 1000 A.D. From 1500 to 1670, the valley remained unoccupied except for hunting parties from the Iroquois of upper New York. Then in a short period, it was claimed by the following nations: France, England, United States, Shawnee, Delaware, Iroquois, and Susquehannock. In the early days of colonization it was the road for British armies, such General Braddock's, to march west, and for Forbes's successful march to take Fort Duquesne and establish Pittsburgh. The valley allowed for a transition from the Allegheny Mountains to the flood plains of the Monongahela River, and then on to Pittsburgh. In 1884, following the natural contours favored by the Native Americans, the Pennsylvania Railroad was built through the area with a small wooden station at the future site of Wilmerding.

Prior to 1890 it was wooded farmland in the deep valley of Turtle Creek, a major contributory to the Monongahela a River, about 15 miles from Pittsburgh. Turtle Creek was a deep and swift creek known for its yellowish to brown color. John Frazier had been the area's first settler, setting up a blacksmith shop and trading post in 1752, where Turtle Creek enters the Monongahela. The area's first visitor would be a young George Washington who spent the night at Frazier's post on his way to explore the French forts of the area. Washington even preferred the location to that of present day Pittsburgh for a major area fort. Washington would visit the valley a number of times thereafter, the last being the officer's first major defeat at the Battle of Braddock. Washington's journal noted Dower's Tavern at Turtle Creek as a recommended stop for rest and dining. Turtle Creek had been a key stop for Indian trails for centuries, and became part of the first British road system in the area. Travelers and marching armies found Turtle Creek to be the ideal last stop before arriving at Pittsburgh.

The inn at Turtle Creek was a relay station on the Pittsburgh and Philadelphia stagecoach line. The creek and hills of Turtle Creek presented a dangerous passage in the early 1800s. A stagecoach traveler in 1810 described the hills as "awful," and the creek swift and deep enough to drown horses. As a key stagecoach stop and river stop on the Monongahela River, Turtle Creek had one of the earliest post offices west of the Alleghenies. The Pennsylvania railroad opened for business with its tracks going from Pittsburgh to Turtle Creek in 1851. In 1859, the Pittsburgh and Connellsville Railroad was extended to Turtle Creek (Port Perry) allowing stagecoach travelers to make daily rail connections to Pittsburgh. Its role as part of a transportation net-

54 I. E. Levine, *Inventive Wizard: George Westinghouse* (New York: Julian Messner, 1962), 134

work continues to this day. Railroad building started in the 1870s following branches of the old stagecoach route. From 1753 to today, the area went from a node on major Indian trails, to stagecoach stop, to river stop, and finally to a stop on the Pennsylvania railroad.

The sleepy rural valley had been visited since the late 1700s by treasure hunters. The legend suggests that Tony Lucas, paymaster of General Braddock's 1755 expedition, buried gold and other coin in the hills surrounding Turtle Creek. The *Wilmerding News* noted in 1904, "many people have expended time and treasure searching for his wealth, but to no avail. History fails to substantiate the story, as it is on record that the French secured the money chests and even Braddock's private papers."[55] Westinghouse, however, added to the legend in 1903 by naming his boarding house in Wilmerding for young employees, the Tonnaleuka Club. Tonnaleuka being the Indian version of the name Tony Lucas.

It was as a stop on the Pennsylvania railroad that George Westinghouse first saw this scenic valley. Interestingly, this little station was the Pennsylvania Railroad's first "Flag Top" station, where switch and block signaling were used. Earlier travelers, such as General Lafayette, Charles Dickens, and Stephen Foster, had noted the beauty of the valley. Turtle Creek Valley cut the Allegheny plateau deeply, a sign of its geological vitality. Floods even today remain an attribute of Turtle Creek Valley. Turtle Creek cut a very fertile bottom, making it excellent for Indians to grow corn and later provided outstanding farmland for settlers. The area had been extremely rich in wildlife as well. It was in the nearby woods that James Audubon made his famous painting of a passenger pigeon. The wooded hills provided habitat for a variety of mammals, including bears, and prior to 1800 even the wood buffalo was common in the area. The rich bottomland offered settlers a new crop —Monongahela rye offered an excellent whiskey making mash in the 1790s. Monongahela Rye whiskey became world famous for decades being shipped down the Monongahela to the Ohio and ultimately to New Orleans, New York, and Europe. These Scottish-Irish settler distillers, eventually, moved west to avoid taxes, and their recipes became the basis for today's Kentucky Bourbon. The great Oaks and Hickories of the area tarnished a barge building industry in raw material as well. Lewis and Clark's riverboat was built in the area. The beauty of the valley had lost nothing when Westinghouse made many trips through the valley on the Pennsylvania Railroad in the 1880s.

In the 1880s, Turtle Creek Valley remained a rural setting as its neighbors, such as Braddock and East McKeesport boomed as steelmaking towns. Turtle Creek valley was described as farmland with only a few log houses in sight. The town of Braddock, less than two miles away, had taken its place with Liverpool, Birmingham, Manchester, Coalbrookdale, Essen, and Sheffield as the world's industrial forges. These mill towns were hailed as economic engines by industrialists and loathed by artists and writers such as Charles

55 *Wilmerding News*, September 2, 1904

Dickens. Philip James de Loutherbourg's 1801 famous painting, *Coalbrookdale by Night*, decried the destruction of the rural British countryside. Inspired by visions such as de Loutherbourg's painting, poet William Blake wrote, "And was Jerusalem built here, among these dark Satanic Mills?" Charles Dickens had traveled through the beauty of the Turtle Creek valley by stagecoach on his way to Pittsburgh. Arriving at industrial Pittsburgh, he compared the Pittsburgh of 1842 to the smoke filled city of Birmingham. These smoky scenes of the early 1800s were nothing compared to the dark daylight hours of 1890 Braddock and Homestead. These sulfur cloaked towns a few miles downriver in 1890 were much closer to the satanic visions of Blake.

Homestead and Braddock offered examples of company towns left to grow on their own. Andrew Carnegie's first steel mill, Edgar Thomson Works, had been located in nearby Braddock since 1875. The town had experienced rapid growth and massive migration to support the mill, which by 1890 was the largest steel mill in the world. The immigrant laborers lived in poor rented one-room apartments. Often whole families lived in single rooms with no running water. Toilets were in communal outside courtyards. Saloons and bars out numbered grocery stores and retail clothing outlets. Drinking helped the workers avoid the reality of the their conditions, but took a huge chunk of potential productivity away from the company. Industrial accidents were a daily experience, leaving as significant part of the population disabled and unemployable. Though part of the population was supported by a strong family structure, but this generosity often doomed the family to stay at the level of poverty. The industrial world of Charles Dickens had come to America in these mill towns. The suffering of his fellow man always moved Westinghouse.

From the very origins of his manufacturing enterprise, he had demonstrated the size of his heart. His primitive 15-man operation in 1870 on the north side of Pittsburgh had started the practice of closing the shop Saturday noon to Monday morning; such a practice was unheard of. He initiated a thanksgiving dinner ritual for his employees and families at the best Pittsburgh hotels. When the company got too large for such a dinner, he became the first to start the practice of giving his employees holiday turkeys. He built parks and baseball fields for his employees. The generosity and big heart of Westinghouse was well known throughout the Pittsburgh district. Westinghouse developed a company funded pension and disability program for his employees, sixty years before most American industries had even started consideration for such programs. Unlike his paternal industrialist neighbor, Andrew Carnegie, his concern for the employees seemed more based on his love of his fellow man than any potential productivity gain. In fact, bankers would point to his generosity as the root cause of the failure of Westinghouse Electric in 1907 (even though, as we have seen, this was more related to banking manipulation).

On trips to England to study railroad signals, Westinghouse observed first hand the industrial slums of Dickens. He had read Jules Verne's *Begum's*

Millions, the story of a dark and dismal steel town that hauntingly resembled the Pittsburgh area. He had also followed with interest the earlier writings of Robert Owen who had designed the model industrial community in New Lanark, Scotland. Owen had caused much discussion in the industrial world with his early success at New Lanark. The New Lanark community had been from the start a planned community for the worker. Owen had become a reformer and preacher, but the operation was profitable, which got the attention of industrialists around the world. The American Management Association today considers Owen's 1813 address to manufacturers a classic in management. The famous quote from that address being:

> Your living machines may be easily trained and directed to procure a large increase of pecuniary gain. Money spent on employees might give 50 to 100 per cent return as opposed to a 15 per cent return on machinery. The economy of living machinery is to keep it neat and clean, treat it with kindness that its mental movements might not experience too much irritating friction.[56]

From 1821 to 1840 Owen's success caused a great deal of interest, but few tried to emulate him except on a piece meal basis. When Westinghouse visited Europe in the 1870s, there were several manufacturing communities in Denmark and Sweden, but there was a closer example a few miles downriver from Pittsburgh. Old Economy, Pennsylvania had become the final settlement of the Harmonists. The Harmonists were German separatist immigrants that had established communal manufacturing centers in the mode of Robert Owen. In fact, Robert Owen was so impressed that he purchased their community in New Harmony, Indiana. Many of the Harmonists returned to Pennsylvania to establish another manufacturing community near today's Ambridge, Pennsylvania. The Duke of Saxe-Weimar visited the community in 1826 and reported:

> Their factories and workshops are warmed during the winter by means of pipes connected with the steam engine. All the workmen and especially the females, had very healthy complexions, and moved me deeply by the warm-hearted friendliness with which they saluted the elder Rapp. I was also much gratified to see vessels containing fresh-scented flowers standing on all the machines. The neatness which universally reigns is in every respect worthy of praise.[57]

The success of these communities was restricted by their unusual religious views, which included celibacy and a belief in the second coming. The community was well past its peak in 1870, but Westinghouse knew of its success, in that they remained, as a community a major stockholder in the Pennsylvania Railroad. Many industrialists including Andrew Carnegie borrowed or modified various practices of these industrial villages. Most of these industrialists approached them from a pragmatic business viewpoint. Westinghouse, however, tended to view the phenomenon as part of a capi-

56 Harwood Merrill, ed., *Classics in Management* (New York: American Management Association, 1960) p. 23

57 Francis Leupp, *George Westinghouse* (Boston: Little, Brown, and Company, 1918)

talistic Christian imperative. These Puritan principles came from his parents in New England. Westinghouse viewed business and workplace fairness as integrated, not motivational. God would bless a business if it followed scriptural principles. Furthermore, the Westinghouse approach was a quite simple application of Christian principles without fanfare or expectations of productivity gains. Westinghouse's dream represented a new model, one compatible with a pluralistic, democratic society. In fact, Westinghouse's Wilmerding leaned closer to socialism than capitalism with his views of a worker community. In 1904, the *Pittsburgh Sunday Leader*[58] credited it with being "the leading center of socialism of the peaceful, sane variety." Clearly, Westinghouse's design naturally favored socialism. From 1902 to 1912, the Socialist party ranked second behind the Republican Party with the Democrats running third. None of this was in any way related to Westinghouse's beliefs, which opposed socialism.

Westinghouse believed a well-designed working community was an extension of capitalism, not socialism. His idea integrated an employee community into American society, not as a separatist organization but a seamless attachment. While Max Weber gets credit for the phrase "Protestant Work Ethic," it was Westinghouse who forged its industrial application. Like industrialist philosopher Robert Owen, Westinghouse saw employee living conditions and corporate productivity as interlinked. Westinghouse had been appalled at the development of the Pittsburgh area. European immigrants had flooded the area starting in the 1870s. No type of urban planning existed, and the immigrants formed slums around the great steel mills of the district. The rapid development left these slums devoid of basic comforts, such as running water, private toilets, sewage management, and reasonable living space. 8 to 20 family members and friends crowded into rented rooms. Alcoholism and disease were common in these slums, which resulted in absenteeism as high as 10%. Furthermore, the twelve-hour day took its toll on family life. The unstructured development fostered a type of racism, which favored clustering of the various nationalities and religions. As early immigrants, such as the Irish and Germans, improved their lot, they became fearful of the newer immigrants, such as the Slavs, Italians, Hungarians, and blacks. This created a type of caste system based on nationalities. Schools were lacking in most towns and illiteracy was widespread.

The mill town of Braddock stood a few miles away, and was typical of area industrial towns in the 1880s. The sun rose at 10am and set at 2pm due to the thick smoke of the steel mills. An afternoon thunderstorm could make things as dark as night. The streets were muddy or dusty depending on the weather. The workers lived in poorly built rented wooden houses, with shared cooking and toilet areas. The neighborhoods were part poor farm and part urban in nature. Pigs and chickens were raised for food, and liquor distilled for relief. Saloons were open twenty hours to accommodate

58 The spelling of Pittsburgh without the "h" occurred between 1895 and 1910

the mill hours and workers thirst. Employees worked 12-hour days, seven days a week, and if that wasn't bad enough they rotated shifts every two weeks using a twenty-hour long turn to make the change. The mill created the social schedule for weddings, holidays, birthdays; even funerals were planned around the mill's pace. Funerals, for example, were in the home, and the women would cook two large meals every twelve hours to accommodate workers coming off their shifts to pay their respects. Sadly, this was a common event, with fatal industrial accidents being a weekly occurrence. John Fitch, a sociologist, who studied steelworkers at the turn of the century in Pittsburgh, found that at age thirty-five there was a perceptible decline in strength. The decline in strength ultimately led to discharge.[59] Unfortunately, few men reached the age of forty without being seriously injured. Fatal accidents occurred at a pace that is hard to comprehend by today's standards. A one-year study of Pittsburgh's Allegheny County in 1907 recorded 526 fatal accidents.[60] Westinghouse had not grown up in such an industrial setting, and its existence appalled him.

After 1886, Westinghouse's many industries advanced from linear to exponential growth with thousands of new employees being added. The physical location on the North side of Pittsburgh needed added capacity to keep up with air brake demand from around the world. Like the Carnegie mills of Pittsburgh, Westinghouse's factories were a magnet for immigrants from Austria, Ireland, Germany, Poland and Italy. Slavs and Hungarians were also coming in larger numbers to take the unskilled jobs. Blacks and Mexicans represented the smallest influx, but by the 1890s were replacing Slavs, Italians, and Hungarians in the lowest industrial jobs. Westinghouse viewed these workers as disadvantaged souls in society and, because they were Catholics, in spirituality as well, although Westinghouse showed no interest in politically or religiously converting these workers. Westinghouse's goal was to help make solid constructive citizens of these immigrants. Westinghouse wanted a planned community to help his workers perform well at work and become owners in the region.

By 1888, Westinghouse was planning a move for Westinghouse Air Brake from Pittsburgh to a new location. He hired architects to help locate a site that could support both the works and a planned community. Westinghouse's business associate Robert Pitcairn recommended the Turtle Creek Valley. Pitcairn was an executive of the Pennsylvania Railroad and a board member of Westinghouse Electric Company. Today the town of Pitcairn neighbors the town of Wilmerding. The Turtle Creek valley offered water, railroad connections, a river port, and availability. The plant and community would be integrated. The factory was designed to be a safe environment with good lighting and ventilation. A clean and well-planned community would

59 William Serrin, *Homestead; The Glory and Tragedy of an American Steel Town* (Random House: United States, 1992), 62

60 Howard Harris, ed., *Keystone of Democracy: A History of Pennsylvania Workers* (Pennsylvania Historical and Museum Commission: Harrisburg, 1999), 129

be built into the surrounding hills. Of particular interest to Westinghouse was the use of parks and greens to achieve a campus like atmosphere. The planned employee houses were to have an emphasis on lawns and trees. Westinghouse's plan would go further than even Robert Owen's New Lanark in Scotland, in that his plan would include all phases of the community. Yet, it would be unique from the American societal imitations of New Lanark (such as Economy, Pennsylvania; Zoar, Ohio; and Harmony, Indiana) in that it would have political and religious freedom. Westinghouse had actually started to offer low cost "cottages" to his workers at Union Switch and Signal in Swissvale around 1886.[61]

Westinghouse's plan seemed more related to ideas of Dr. Benjamin Ward Richardson in the mid-1800s than those of religious industrial organizations. Dr. Richardson was known for his studies on the impact of industrial environment on tuberculosis, alcoholism, and nicotine addiction. Jules Verne made his ideas on the planning and design of worker communities famous in 1879 with the publication of *The Begum's Million.* Richardson's design details were very close to those in Wilmerding. For example, Richardson called for "Every house to be on a lot planted with trees, lawn, and flowers," and "houses thirty meters from the street." He emphasized flowers whenever possible, prescribed ventilation requirements, plumbing, and drainage requirements. Westinghouse set similar requirements on construction. He planned in citywide parks and gardens using expert gardeners to maintain them. It was, however, never his plan to control, but rather to inspire and create a culture. It was a model emulated by fellow Pittsburgher, Andrew Carnegie, who built libraries but required the communities to supply the books. As to the individual homes, he offered prizes, rewards, and awards to assure the beauty of the workers' homes. He set an example with the gardens surrounding the town's Westinghouse headquarters building known as the "castle." He created a culture of beauty that lasted over hundred years, the remnants of which can be seen even today.

The center and heart of the town was Westinghouse Air Brake, and he built the town to compliment the factory. The factory in 1890 became known as the largest machine shop in the world. Unlike the famed industrial factories of the time, Westinghouse planned a safe and comfortable workplace. He paid special attention to restrooms, lunchrooms, and rest areas. Women employees had special lunchrooms with tablecloths. Artesian wells were dug to supply safe drinking water for the employees. The cleanness of the plant was emphasized and maintained to strict standards. Lighting was also state of the art throughout the plant. A number of these features were directly related to experiences at his father's New England manufacturing plant. Westinghouse Air Brake headquarters was a beautiful castle, modeled after one in Scotland. The original central house, however, was a wooden frame house that served as a community club with a library, bowling alley, and swimming

61 "Building Intelligence," *Manufacturer and Builder*, May, 1887

pool. It was very similar in scope to Carnegie's first library in Braddock few miles away with its bowling alleys and swimming pool built about the same time. One big difference was that the Westinghouse employees had time to enjoy it, while Carnegie's steelworkers were working 12-hour shifts with a day off every few weeks.

In 1893 the Railroad Safety Appliance Act was passed by Congress requiring the air brake and automatic coupler on all trains. Lorenzo Coffin who had called for the first trials in Burlington, Iowa created the bill with the help of Ohio senator, William McKinley. The bill would create a new prosperity and an influx of European immigrants. Later in the 1890s, Westinghouse created a partnership with the Young Men's Christian Association (Y.M.C.A) another idea he had discussed earlier with his friend William McKinley. From 1896, the Y.M.C.A shared the "castle" with headquarters for Westinghouse Air Brake. For years the main role of the Wilmerding Y.M.C.A. was to integrate foreigners into American society. New immigrants in Wilmerding were sent notices of meetings for citizenship, which were conducted by lawyers. Beginning emphasis was on English classes, in which the association worked with the public schools. In addition, the association worked with the Daughters of the American Revolution to present extensive American history programs prepared in seven languages. This cooperative networking was fundamental to the character of the town. In the summer months, the park was used for educational moving pictures. A stereopticon was used as a projection camera. Various languages were used interchangeably in the lectures, which covered the prevention of infectious diseases, prevention of tuberculosis, and American history. This foreign program became the model for the whole United States. The Y.M.C.A. shared facilities of the castle with Westinghouse Air Brake management until 1907. In 1907, Westinghouse completed the "Welfare Building" for use by the Y.M.C.A., and shortly after he built a building for the women (Y.W.C.A). These had world-class swimming pools and gymnasiums. The Wilmerding Y.M.C.A became the second largest in Pennsylvania by 1910. It was available to all in Turtle Creek Valley.

At the castle, further educational opportunities were offered. These included basic math, electricity, grammar, typewriting, machine design, among others. One specific course focused on the construction and operation of the air brake. Machining and technical training was essential to Air Brake because the available immigrant workforce consisted mainly of unskilled European peasants. Westinghouse needed to teach them English even prior to technical training. Boy apprentices were paid an hourly rate as they attended night school courses. The night school became known as the Casino Night School. Graduates of the school tended to have a fast track in the company, and most of the plant management were graduates. In 1904, it was reported that 160 students were taking evening courses from nine instructors. Besides the courses, there were a wide variety of special lectures and talks. Westinghouse managers were expected to be instructors and lecturers. Even his plant managers spent time in the training rooms. Many clubs were organized including chess and checkers. The physical programs offered would rival today's health clubs. Special classes were offered in tennis, wrestling, and fencing, all favorites of George Westinghouse. In addition, the association hired a prominent expert in "physical culture institution," C.H. Burkhardt. Burkhardt built and developed an outstanding track and field program for the town. Services were also held on Sundays and evening Bible schools were offered. These services were a bit problematic with the Roman Catholics fearful of protestant indoctrination. It is interesting that even today the "castle" dominates the town of Wilmerding, representing the symbiotic relationship of the company and community, while nearby steel towns such as Braddock and Homestead were dominated by the mill Superintendent's house or the Carnegie Library, symbols of company power.

Westinghouse's approach to employee housing was much different than that of others. He had experimented with low cost employee houses in Swissvale. These Swissvale "cottages" were six-room homes costing around $1200 to $2000. Westinghouse wanted to build a better house for his Wilmerding project. It focused on true ownership by the employee. Initially, in the 1890s, Westinghouse Air Brake purchased the land and built individual houses. The houses were then sold to the employees at cost. The mortgages were adjusted to the employee income and allowed for payments over fifteen years. The houses were also covered by a type of mutual insurance that protected the owner in times of unemployment, disability, and death. This stood in stark contrast to the company mining towns, in which, when a death occurred, the company quickly foreclosed on the house. Furthermore, Air Brake's maintenance department would help repair utilities and appliances, as well as some major repairs. In general, the company assured that all houses were maintained in a state of excellent repair. The housing plan was altered by 1902 as reported in the *Wilmerding News:*

> The purchaser of any property is required to pay about one-fifth of the purchase money in cash upon delivery of deed. He then executes a purchase-money mortgage, payable in five years, with interest payable quarterly at the rate of 5 per cent per annum. While no requirement is made, it is expected that the purchaser shall reduce the principal of the mortgage quarterly by such payments on account, as he may be able to make. This plan enables him, during hard times to keep the transaction in good shape by merely paying the interest, while on the other hand, when good wages are earned, he can discharge such a part of the principal of his mortgage as he may desire.[62]

This type of mutual investment by the company and the employee was characteristic of Westinghouse. He believed in this type of giving versus the giving of Carnegie and others. This philosophy made for a very different type of "company town." Westinghouse always stood as a lender or backer of last resort for any of his employees. He could not and would not act as a banker of the times with coldness. His heart was too big and too Chris-

62 *The Wilmerding News*, November 23, 1904

tian. He was known to help employees with all kinds of personal problems. Westinghouse proved that capitalism could be compassionate in the time of the "Robber Barons." Westinghouse was not a communist, which many of the time believed, partly because the town of Wilmerding seemed to attract many socialists. Westinghouse's approach was unique, but he had no socialist or communist leanings, only a passion for the team that made him a success. He differed from communal manufacturers, such as Robert Owen, in that; there was a mutual investment of employees and company, but the main difference was that Westinghouse's model was adaptable to heavy industrialization. Few of his industrial counterparts embraced his approach.

Westinghouse tried to keep his social programs and civic efforts as separate as possible, but he was always ready to intervene to fill needs or assure living quality. One such example was the Tonnaleuka club. Wilmerding had a growing population of professional young men in the engineering and technical trades. The local private hotel failed to supply comfortable living quarters for these men. In 1901, Westinghouse purchased the Glen Hotel building on Marguerite Avenue. Westinghouse reworked the building expanding and adding living rooms and dining rooms. In addition, there were smoking rooms, a billiard room, bowling alley, and a library. A club was formed so that about 30 young men could enjoy this higher class of temporary living; additionally an associate type membership was developed for others to enjoy the excellent recreation rooms and fellowship. The Club, in particular, catered to single professionals moving to Wilmerding. The success of the Tonnaleuka Club in Wilmerding was duplicated a few years later with the "East Pittsburgh Club," which serviced employees of Westinghouse Electric in Turtle Creek and East Pittsburgh.

Andrew Carnegie, in particular, saw Westinghouse's views as too close to socialism, yet they shared a similar belief system. Fundamental to both men's views was the idea that you need to help those who help themselves. Both men believed in competition and capitalism, and neither believed in unions. Westinghouse was a paternal capitalist, but in a different vein than Andrew Carnegie. Carnegie believed that he and certain others were destined to be public trustees of capital. Carnegie was more patriarchal than paternal. Westinghouse was more like the father that taught his sons by buying them tools and training them how to use them. He was a follower of fellow New Englander, Ralph Emerson, in his insistence on self-reliance. Westinghouse gave freely, demanding nothing in return. This is where he differed from similar programs developed later by Henry Ford. Ford would however, demand a certain type of moral code from his employees in return. Westinghouse and Ford both, however, believed in manufacturing assets over capital assets, whereas Carnegie believed in the primacy of capital in manufacturing. This emphasis on manufacturing assets naturally extended to the employee. The other characteristic of Westinghouse employee programs was the planning for bad times. It was a personal philosophy of Westinghouse, which he had applied to all of his companies, with the notable exception of Westinghouse Electric, which he leveraged to support major engineering projects.

The town of Wilmerding was formed as a separate political entity, but Westinghouse often allowed company resources to be used for the enhancement of the community. The company employees often maintained the parks and streets, although the borough did pay reduced rates for the company services. The borough was legally incorporated in the state of Pennsylvania, and it functioned similarly, politically to that of neighboring borough. The town did tax houses at a rate of 10 mills to support the schools. This tax rate was about average for Pittsburgh area boroughs. The borough taxed businesses as well. It also, on a very limited basis, floated bond offerings. Schools were part of the public school system of Pennsylvania. The borough paid for street paving and street lighting, but again the company would help when necessary to maintain the beauty of the town. Wilmerding was always uniquely a Westinghouse Air Brake town with over 85% of the residents (the balance being small businesses and services) being Westinghouse employees.

Another unique feature of the borough of Wilmerding was its economic and social mix. While still a predominantly a "mill town," it had a large segment of college graduates working as managers, engineers, designers, department heads, scientists, and administrators. Westinghouse was almost the exception in that he lived in Homewood, and took the train to Wilmerding each day. Wilmerding offered a refreshing oasis in the smoky and dirty boroughs of the steel companies. Politically, the majority party was Republican as was Westinghouse himself. It may seem odd to today's reader that the immigrant workers of Pittsburgh were solidly Republican, as it did to me that my grandfather, a Pittsburgh steelworker was a strict Republican. The Republicans were the party of Lincoln and trust-busting Teddy Roosevelt, both of whom were extremely popular with the immigrant laborers and native Scotch-Irish Jeffersonians of the area. Republicans dominated the mill towns of Pittsburgh until the New Deal of the 1930s when the Democratic Party took over. The Democratic Party ranked only third with the Socialists ranking second (by ballots cast), but it was a minority of socialists that distinguished the borough from nearby mill towns. The Socialist Party probably got more press than it deserved, primarily because some elected borough leaders were Socialists. It is doubtful that registered Socialists ever exceeded 3% of the population. Its roots in Wilmerding had nothing to do with George Westinghouse. Westinghouse, a staunch Republican, had no sympathy, yet alone empathy for the socialists. In 1905, the population of Wilmerding was about 5,000. The town, however, was part of a Westinghouse dominated Turtle Creek Valley consisting of East Pittsburgh, Turtle Creek, Pitcairn, and East McKeesport. In particular, East Pittsburgh was a Westinghouse Electric town. The valley population was 30,000 in 1905. Politically, efforts to combine these communities into a "Westinghouse Valley" failed due to strong resistance of the town of Wilmerding, which maintained a superior position in services, educational resources, civic pride, and beauty. The first movement to combine Turtle Creek, East Pittsburgh, and Wilmerding in 1900 seemed logical since they were home to basically Westinghouse employees. The idea was to call the combination Westinghouse or Westinghouse City, but Westinghouse was opposed to this.

The borough of Wilmerding fostered superior services and utilities for its residents. The five-man police force was actually larger than that of boroughs of equivalent size. The volunteer fire department was considered one of the best in the valley. The volunteers were paid by the hour when called out by the borough. Westinghouse donated fire trucks. Wilmerding had not only paved streets but also paved sidewalks before most of the Pittsburgh area towns. As noted, all houses had running water (a rarity in some boroughs) supplied by the Pennsylvania Water Company. Because of complaints regarding service and high electrical rates of the Monongahela Light, Heat & Power Company, the town formed its own independent illuminating company. In 1904, using local capital, the United Electric Light Company was formed to supply electricity. As noted, the Pennsylvania Railroad connected Wilmerding to Pittsburgh from the beginning. By 1893, electric streetcars connected Wilmerding to the major towns of Braddock and McKeesport. Wilmerding residents used the streetcars to shop for the harder to find items in Braddock and McKeesport.

Another distinction of Wilmerding among the mill towns of Pittsburgh was its cultural and social attributes. The racial and nationality mix of Wilmerding was similar to near-by mill towns. Westinghouse had shown no bias whatsoever in his hiring practices. In 1904, seven Christian churches including Roman Catholic, Methodist Episcopal, Presbyterian, United Presbyterian, Episcopalian, Lutheran, and United Brethren were represented in Wilmerding. The Roman Catholic was the largest denomination. What was different was the lack of taverns and bars so dominant in the other mill towns, as well as the prevalence of gambling. This was due to culture versus any legal or religiously imposed morals. Alcoholism, which plagued near-by mill towns as a means of escape found less fertile soil in Wilmerding. Clearly, the impact of the Y.M.C.A and the churches had impacted the community. Just as important was the example of Westinghouse, whose behavior inspired a corporate and community culture. While ethnic lodges and clubs existed their role was necessary because the community as a whole offered educational, social, and welfare services. Wilmerding had often been described as a "Midwestern college town." It offered a clean and very wholesome environment in a time of industrial plight.

The Air Brake factory was as revolutionary as the town. The factory would be one of the safest plants built with ventilation systems, sanitary restrooms, and state of the art lighting. Westinghouse developed an employee committee system to take ownership of safety to the employees. The plant had its own Fire Chief and Safety Chief. These chiefs had the authority to shut down operations if necessary to assure safety. The Safety Chief had a committee of eight employees. Each committeeman had a section of the plant, but there were also planned audits of other areas. Committeemen held weekly meetings and discussed areas needing improvement. Safety committeemen received extra pay for the work required. Westinghouse Air Brake had a number of employee improvement programs including improvement teams. Employee improvement at Air Brake was years ahead of this time.

Just as innovative was the plant design, layout, and operation. Westinghouse's automation augured the assembly lines of the 1910s. Westinghouse had spent a year at the start of his career studying metallurgy and the casting process. He had pioneered and improved the steel casting process. Air Brake would in 1890 open the world's most automated foundry to cast brake parts. The foundry while associated with Air Brake would make castings for Westinghouse Electric and Manufacturing and Westinghouse Machine Company. The new foundry used conveyors to move the molds for casting throughout the foundry. He employed the earliest known machinery to make the sand molds, which normally was a highly labor intensive operation. Conveyors then moved the molds to be filled with hot metal from furnaces known as cupolas. The conveyer then moved the mold to a shake out area to remove the castings. The sand was then recycled by conveyor back to the molding area. The conveyors were balanced to achieve a continuous operation twenty years before Henry Ford applied the principle to auto manufacture. In addition, he built specialized equipment for scraping and cleaning the casting. The foundry would host a stream of visiting engineers and executives daily from all over the world.

CHAPTER 12. NIAGARA'S WHITE COAL

With the building of Wilmerding, the new air brake works, and his triumph at the fair Westinghouse was approaching the pinnacle of his career. The Fair had been a huge project, but Niagara would dwarf it in size and scope. Westinghouse had always looked at the Chicago Fair as a necessary step to harness Niagara Falls. The technical community was moving to Westinghouse's camp, but his main competitor General Electric was morphing into an AC electric company. Westinghouse already had hydroelectric plants at Portland, Oregon, and Telluride, Colorado, but these were small operations. These were shorter transmission distances. In California, he had just completed an AC power station 28 miles to San Bernardino. In addition, he was now in debt in the millions in the middle of a national depression. If things weren't bad enough, Westinghouse was a common discussion topic in J. P. Morgan's library office. Morgan well remembered Tesla's 1890 prediction of AC power, "to place 100,000 hp on a wire and send it 450 miles in one direction to New York City. . . and 500 miles in the other direction to Chicago."[63] Morgan always looked to the bottom line. Morgan and Westinghouse did share the common desire to achieve. Niagara offered both the capitalist and the engineer an achievement on the level of a wonder of the world. With its tunnel and generating plant it would be the greatest and most costly engineering project ever seen.

63 Frank Stetson, *Memorabilia of William Birch Rankine* (Niagara Falls: Power City Press, 1926), 28

Only one natural wonder could offer such an opportunity to achieve as Niagara Falls. *Harpers Weekly* said in 1889, "There are some things that do not bear description, Niagara Falls and Mark Twain are two of those things." For the American Indian the power of Niagara was the embodiment of the Great Spirit. Indians would annually offer a canoe with their fairest maiden in it to the Great Spirit. Niagara Falls was the number one tourist attraction in America; even Thomas Edison had honeymooned there. Its power was obvious to the earliest settlers. The power of the falls has an almost unbelievable flow of 100,000 cubic feet per second with a peak of 225,000 cubic feet per second. There is a peak period of flow from April to November. There is even a daily variation of flow, with 9am being the peak and 9pm the low. German electrical engineer, Professor Siemens estimated in1888 that, "the amount of water falling over Niagara is equal to 100,000,000 tons [of coal] an hour."[64] Sir William Siemens calculated that "if steam boilers could be erected vast enough to exhaust daily the whole coal output of the earth, the steam generated would barely suffice to pump back again the water flowing over Niagara Falls." The power of waterfalls through out the world remains preeminent in electrical generation even today accounting for 25% of all power generated. The earliest efforts to obtain power from the Niagara River used waterwheels to power grist mills, flour mills, and saw mills in the 1810s.

Augustus Porter suggested in the 1830s a canal that would bypass the falls, but divert water in a type of raceway down to the bottom of the gorge. This 'hydraulic" canal would have water turbines connected by belts to nearby machinery. The use of water turbines for power was growing in popularity on the Erie Canal. George Westinghouse studied Porter's proposal closely feeling the water turbine could be used for his electrical generation. The Niagara Falls Hydraulic Power and Manufacturing Company was formed in 1853. Construction on the canal started in 1860, but the power system was not completed until 1873 after seventeen years of effort. The operation went bankrupt in 1877, having only one customer, a flourmill, even though it was the largest in the world. Jacob Schoellkopf purchased the canal the same year. He used a system of belts and drive shafts to drive machinery. In 1881 a small company took over the canal and generated a small amount of DC power. It was a tourist attraction, but was forced into bankruptcy in 1883. Schoellkopf became interested in generating enough electricity to illuminate the Falls. In 1881 the Falls were illuminated on special occasions by chemical lime (burning calcium) lights as were theaters of the time, which is the origin of "being in the limelight." Later that year Schoellkopf teamed up with Charles Brush to light the Falls with sixteen arc lamps from a DC generator driven by canal water. Schoellkopf had talked to Westinghouse in 1880s who suggested AC for illumination, but lacked any capability of supplying a full system.

For the story of Niagara Falls we need to go back to the late 1880s. Thomas Evershed in 1886 submitted a plan for electrical generation using a

64 *Manufacturer and Builder*, November, 1888, Volume 20, Issue 11

2.5-mile water tunnel. He proposed DC power, and couldn't find investors because like the belt drives the power availability would be around a half mile. A $100,000 prize was offered for ideas to extend the power transmission over twenty miles; there were no responses. Niagara had started out as a local project, but capital could not be raised, moving the project to the banking circles of New York. A young New York lawyer, William Rankine, was part of an initial company called Niagara Falls Power Company, which had failed to raise the large amount of capital needed. Rankine was a debonair socialite, who had developed a powerful network of attorneys and bankers. Rankine went to a New York lawyer, Francis L. Stetson, who had many tycoon friends. Stetson was considered one of America's most powerful lawyers with two famous clients —Vanderbilt and Morgan. Stetson had played the role of "power behind the throne" in the formation of the great Morgan railroad trust. Stetson and Rankine also started to secure land near the Niagara River. J. P. Morgan got involved by autumn 1889, helped organize an exploratory group of Mr. Jacob Astor, Mr. Edward D. Adams, Mr. Francis L. Stetson, and Edward A. Wickes. Jacob Astor, being part of America's elite was a powerful banker and friend of Morgan. Similarly, Adams was a friend of Morgan and, like Astor, had helped form the Railroad Trust. This was clearly a Morgan team, and one of considerable power in banking, political, and engineering circles.

Initially, whether hydraulic, pneumatic, or electric processes would harness Niagara was still in debate. In 1888 Adams was the second largest stockholder in Edison Light Company. Subsequently, Morgan formed Cataract Construction Company to be the prime contractor of the project, cataract being the geological name for such a great waterfall. Cataract Construction would be the largest project ever undertaken by New York bankers, with capitalization near $3 million dollars. Morgan had even brought in Westinghouse banker, August Belmont, although in a minor position. Edward Adams would be named President and Doctor Coleman Sellers, chief engineer. William Rankine became Secretary of the Cataract Construction Company. Sellers was a professor of mechanics at the Franklin Institute as well as professor of engineering at Stevens Institute. Sellers immediately studied the geology and difficulty of the building of a water tunnel to harness the waterpower of the Falls. To appear impartial, Adams sold his shares in Edison Light.[65] Still, J. P. Morgan was open to the best application of his money. He was far too smart to let technical battles blind his interest in making money.

To that very end, the Cataract Construction Company formed an International Niagara Commission to look at ways to harness and design options to get energy from Niagara Falls. This international commission was needed because part of the falls was in Canada, and investors from both countries and others would be needed. The banking concerns required would demand the inclusion of the Rothschilds, who eventually partnered with Morgan.

65 Seifer, 134

The world's most famous physicist Lord Kelvin (then known as Sir William Thomson) headed that Commission. Lord Kelvin was also an acclaimed engineer with success as part of Trans-Atlantic Cable. He originally was a die-hard opponent of AC power. In addition, the Commission included Doctor Coleman Sellers of Cataract Construction, Professor M. E. Mascart of the Institute of Paris and the College of France, British physics Professor William Cawthorne, and Theodore Turretini of the Geneva Works Department. Their mission was to look at "a central *hydraulic* power station to be located above the falls and to develop as much power as the section of discharge tunnel, the head water, and the hydraulic slope would permit, and for the transmission and distribution of this power overhead or underground by electricity, compressed air, water, cable, or other means, to a manufacturing district [in Buffalo approximately 20 miles away]." For the New York bankers the cost of this project required a non-biased approach. Edison believed it had to be DC current and Westinghouse favored AC, but the commission was even exploring compressed air, and Westinghouse suggested an alternate system of cables and compressed air for power transmission.[66] A number of European firms submitted plans for compressed air as well.

The Commission faced one of the world's biggest engineering challenges in how to glean the power of Niagara Falls for practical use. The first idea was that of a Napoleonic style scientific contest. These financial awards had stimulated many great scientific advances in Europe such as Bessemer steel, aluminum production, Krupp artillery, and the steam engine. The $3000 offer went out for the best proposal. Westinghouse had two of his own engineers, Louis Stillwell and H. Byllesby, in Europe. They were encouraged to participate, but Westinghouse refused feeling he would be giving too much information away on his system. Westinghouse realized that Morgan men dominated the Commission and his technology would easily find its way back to General Electric. The Commission did receive twenty plans as a result of the contest. Most of the twenty were for compressed air, but there were four DC plans and two AC plans. Only one of the AC plans was a polyphase system, by a Professor George Forbes of Glasgow (Professor Forbes would later become consulting engineer for the Cataract Company). When Forbes was brought before the Commission he made the strongest case stating, "It will be somewhat startling to many, as I confess it was at first to myself, to find as the result of a through and impartial examination of the problem that the only practical solution lies in the adoption of the alternating current generators and motors . . . . The only one is the Tesla motor manufactured by Westinghouse Electric Company which I have myself put through various tests at their works."[67]

66 Edward D. Adams, Niagara Power: 1886-1918 (New York: Niagara Falls Power Co., 1927)
67 Charles Scott, "Nikola Tesla's Achievements in the Electrical Art," *AIEE Transactions*, 1943

Westinghouse had assembled a team of outstanding engineers, Albert Schmid, Louis Stillwell, Benjamin Lamme, Charles Scott, and Oliver Shallenberger with Nikola Tesla as a consultant and friend. This team supplied the technical firepower as well as the administrative punch in the R&D effort. Schmid was a Swiss mechanical engineer, who Westinghouse met in Paris. He had worked on Europe's earliest dynamos and arc lighting systems. Schmid had been the superintendent of the Garrison Alley research lab since its beginning. Louis Stillwell was a graduate engineer and had joined Westinghouse in 1886. He had been a proponent of AC current prior to Tesla coming to Pittsburgh. Stillwell had been the organization's earliest supporter of the Niagara project. He had worked early on as an assistant to Oliver Shallenberger, and he had helped to the develop the AC current meter. Stillwell with Shallenberger had opposed Westinghouse adoption of 60 cycles preferring 133 cycles current. Stillwell, however, played a key role in the success of Westinghouse's Telluride AC hydroelectric plant. Shallenberger had come from the Naval Academy, and was a brilliant experimenter having developed current meters and transformers. Charles Scott was a graduate in electrical engineering from Ohio State. Scott functioned as Tesla's assistant at Westinghouse Electric. Charles Scott, a supporter of Tesla's 60 cycles played a key role in changing Westinghouse's mind. Scott headed up Westinghouse's transformer department. Benjamin Lamme was also an Ohio State electrical engineer. Lamme was the youngest of the group, and the most brilliant. Lamme's work on the rotary converter (AC to DC current) would be key to Westinghouse Electric's success. This great technical team would have to deal not only with technology, but the Morgan businessmen, such as Edward D. Adams.

Few men had the full confidence of Morgan like Adams. Edward Adams had shown flexibility, discernment, and ability to compromise in working with Morgan's railroad trust. Adams represented Morgan well as an adroit businessman. He realized that AC power was gaining acceptance and offered more advantages than DC. He conferred with Morgan almost daily by telegram. Morgan, Adams, and even General Electric's Coffin realized that Westinghouse had too powerful a position in AC power generation. In Europe, Adams had been visiting the Swiss power stations and had developed a friendship with Charles E. L. Brown who was Europe's greatest AC engineer. Brown was a proponent of polyphase AC power, like Westinghouse. Brown was a Swiss born engineer and the son of a steam engine manufacturer. He was Director of Operations for Oerlikon Company that had extensive experience in AC power generation. His technical papers in 1891 had stirred the engineering use of AC power in Europe. In 1892 he formed a partnership with Michael von Dolivo-Dobrowolski, who had the competing claim and patent for Tesla's AC system. Probably with some help of J. P. Morgan they found money to form Brown, Boveri, and Company. This company could form opposition legally to Westinghouse and Tesla's AC patents. Perhaps more importantly, Adams channeled Brown's technology back to General

Electric. General Electric could then start an AC program of its own even though Westinghouse would challenge it in court. This was typical of the strategy of both General Electric and Westinghouse Electric to push ahead with production and development, while the courts argued the legality. The argument centered on the fact that it was a polyphase AC system, while Westinghouse's system was really two phase (Tesla's actual was polyphase). It should be noted that polyphase did operate more effectively than two phases. The Morgan team did have one important free thinker on the team, which would help keep some balance.

Morgan might have been richer than God, but John Astor III (owner of Waldorf-Astoria Hotel) was richer than Morgan. He may well have been the richest man in the world with assets of $100 million compared to a mere $30 million of Morgan. Astor was a Harvard graduate and only 25 years old when he had become fascinated by astronomy. Astor was an accomplished inventor as well, and had won a prize at the Chicago World Fair for his pneumatic walkway. At the Fair he became a friend of Nikola Tesla. Interestingly, Astor shared a belief in life on Mars as did Tesla (Elihu Thomson and Lord Kelvin), which helped develop an unusual bond between the two inventors. Tesla also became a regular at the Waldorf-Astoria as well as the Players Club and other haunts of Astor. Tesla and Astor became poker friends with Henry Clay Frick, who also stayed at the Waldorf-Astoria. All of these men had come over to the future of AC power.

Thomas Edison had, however, hung a DC millstone around the neck of General Electric, but with Thomas Edison effectively sidelined at General Electric, the company could more forward with AC power generation. In 1892 General Electric used this Brown polyphase AC technology to bid and win the contract for the Redlands, California power plant. The power station was on Mill Creek, a branch of the Santa Ana River in the snow covered Sierras. The project progressed with the company known as Southern California Edison Electric. The biggest customer of the Redlands plant would be Union Ice Company, which would ship 7000 tons of ice every year to Los Angles at a price 50 cents cheaper than local ice plants. While the plant wouldn't come online until November 7, 1893, it would be the first polyphase AC power plant in the world. The progress of this AC project would allow General Electric to have an alternative AC plan in the bid for Niagara. It was now clear to Westinghouse just how aggressive, flexible, and competitive Morgan's General Electric could be with Edison. He also realized that without Tesla's patents, the Morgan electrical trust would have crushed him long ago. Tesla's patents had allowed the chess match between Morgan and Westinghouse to continue. It is interesting that Union Ice Company was using the pioneering process of ice making by Westinghouse, Church, Kerr, and Company.[68] Westinghouse Machine, Union Switch, Air Brake, as well as Westinghouse, Church, Kerr and Company were all developing new

68 "Artificial Cold," *The American Architect and Building News*, August 5, 1893

products related to the increase of electrical power. Westinghouse's vertical integration was far ahead of the market itself, but would give his companies an edge that would remain for a century. The firm of Westinghouse, Church, Kerr and Company advertised the following new markets for their refrigeration system in 1893:

> The field for artificial refrigeration seems to be endless, including markets, creameries, refineries, chemical works, wine-cellars, confectioners, steamers, hotels, hospitals, apartment houses, safe deposit (for the storage of furs), and similar industries in which it has not thus been available, to say nothing of cold-storage warehouses, breweries, packing-houses, etc., in which it has already become an established necessity.

Meanwhile, as the Commission argued the technical designs for Niagara, Cataract Construction moved ahead with the tailrace tunnel at Niagara that would be needed with any design. On October 4, 1890, Cataract broke ground at Niagara. The mile long tunnel was itself a major engineering challenge. 1300 workers were hired for the monumental excavation. Both dynamite and compressed air drills would be required. The drills were powered by 238 wheel wheels along a canal above the Falls (a year later they would switch to electrical power). The mile long horseshoe shaped tunnel would use Niagara water rushing at twenty miles per hour. It became the largest water tunnel in the world and one of the world's great engineering feats. Waterpower would first be converted to mechanical power by turbines at the end of the tunnel. Morgan and most of the New York bankers were firmly committed to the success of the project because of the money invested. In Europe Adams continued to at least buy time for General Electric.

By mid 1891 the situation looked grim for the hopes of the DC option. *Electrical World* reported that America had only 202 DC power stations compared to 987 AC power plants. The support of Lord Kelvin had been the major block to the acceptance of AC. Coleman Sellers on the Commission also helped to politically block it. Professor Forbes joined the commission and Cataract in April of 1892, and his backing of AC would assure its success. Some AC plants had come under General Electric with the merger with Thomson-Houston (formally concluded April, 1892). General Electric could bid on an AC system, but the pending lawsuits over Tesla's patents made it questionable as to their ability to deliver. General Electric held to its DC plan opposing the use of AC. The only reason to oppose AC was the Westinghouse Electric ownership of the Tesla AC patents, but now General Electric was progressing via the Dobrowolksi challenge. General Electric had discovered another key resource that had been in an obscure lab of Thomson-Houston prior to the merger into General Electric. Charles Proteus Steinmetz was a hunchbacked small man, but the mental equivalent of Tesla. He was a mathematical genius that would turn the description of alternating current into a code of mathematical formulas. It was said that only a couple of engineers in the world could fully comprehend his formulas, but in them lay the secrets of AC power. Working with Dobrowolksi notes, he designed a brilliant three-phase AC system. While brilliant and for the first time mathematically defined, it offered nothing new over Tesla's original ideas. General Electric now, however, had a genius to replace Edison and take on Westinghouse and Tesla. Coffin was also building a strong technical organization along the lines of Westinghouse Electric.

The resistance of Lord Kelvin, as well as Edison to AC power gives us insight into the real revolution in science and engineering that was taking place. Lord Kelvin had stated that the best electrical engineer "is a good mechanic who has acquired a smattering of electricity." Teslan AC physics represented both an evolutionary and revolutionary change in physics and engineering. Prior to Teslan physics, Newtonian physics ruled with its hands-on, physical experimentation. Scientific proof of theories was rooted firmly in physical proof. A Newtonian mechanic could come to an understanding of physics and electricity through his own experience and physical experimentation. Like Newton, Edison and Kelvin, were true geniuses in Newtonian applications. Teslan physics changed this to a mathematical approach to theory with physical proof. Teslan physics went even further with the use of imprecise mathematics, such as differential equations, which went beyond simple direct or inverse physical and mathematical relationships. Teslan physics brought an abstraction in science and engineering that bothered practical engineers, Newtonian scientists, and mechanics. I'll never forget my own personal disappointment with my first electrical engineering course at the University of Michigan. I went in hoping to learn the skills of my electrician grandfather and found myself using differential equations and partial solutions for abstract concepts. Teslan physics can be disturbing for the engineer who loves precise physical solutions. Westinghouse noticed this later on in his apprentice programs finding that there were two distinct types of apprentices and they appeared so unique that it was uncommon for one to be comfortable in both group; one group being the practical electrician versed in mechanics, and the other the electrical engineer versed in abstract science and calculations.

With Teslan physics for the first time the engineer could be separated from the laboratory using mathematics to theorize and design. This pragmatic mathematical approach was actually feared by practical scientists such as Edison, who had a hired mathematician to do his calculations. In fact, both Edison Electric and Westinghouse Electric had large "Calculating Departments" to do this mathematical work and theoretical design. Westinghouse was rare in that he could deal with abstract concepts and mechanical reality with extreme ease, but his organization reflected the dichotomy of the electrician and electrical engineer. Westinghouse became the Rosetta stone for the two groups. Tesla himself was an outcast at both Edison and Westinghouse Electric, where electricians dominated. Westinghouse's use of electrical engineers, such as Tesla and Lamme, was in many ways his real genius. Teslan physics would be an evolutionary and intermediary step for Einstein physics, which used mathematics as proofs of theory, waiting decades for physical verification. Westinghouse and his support of Tesla's principles gave him an edge in this period of engineering change. Physical and practical applications of Tesla's concepts, however, started to win over scientists and engineers in the 1890s. This is where the Niagara Commission, as well as General Electric, found themselves in 1893, when surprisingly, Lord Kelvin changed his mind on AC power in early 1893. Westinghouse had won him over with the Telluride mining power, which was built in 1890. Westinghouse biographer, Henry Prout, even credits Telluride with winning over Westinghouse who was still looking at compressed air power transmission at Niagara. Telluride's power plant sent AC current two and three quarter miles to the mine.

On May 6, 1893, it appeared that Westinghouse had finally won as he received a telegram that the Commission had committed to polyphase AC power for Niagara. The decision for AC had hinged on Lord Kelvin switching to AC current. Westinghouse notified Pittsburgh of the decision to cheers from his management team. Adams, however, telegrammed Coffins and General Electric a few days earlier of Lord Kelvin's change of mind. General Electric had correctly anticipated the AC decision having developed its own AC system. Even more importantly, Morgan and his associates had anticipated the need for an AC backup plan. Morgan had covered his bets. General Electric had its own AC system, although Westinghouse contested it. He had backed the formation of a competing AC European company in Brown, Boveri, and Maschinenfabrik Oerlikon, but they were also being contested legally. But just as Westinghouse appeared assured victory at last, a fourth rival appeared on May 11, 1893.

This rival was a Professor George Forbes who had a year earlier been named consulting engineer for Cataract Construction after severing on the original Niagara commission. To some Westinghouse managers the earlier assignment of Forbes to Cataract seemed strange because of his absolute support for AC current and Westinghouse Electric. Why would Morgan do such a thing? It now became clear, Forbes having seen Westinghouse's plans over the past three years as well as having visited Westinghouse's operations as a representative of Cataract was in position to "design" an AC system, and Cataract announced that Forbes would supply the dynamos to Niagara Falls. Morgan couldn't count on a legal victory by General Electric or Boveri so he had hedged his bets yet again. Forbes, while an eminent scientist, needed Morgan's money to build a company to get into the battle. With Forbes, yet another eccentric scientist entered the story. Forbes was an arrogant power seeker, the exact profile for a Morgan operative, and the partnership would bring in millions for both.

It was another unethical, if not illegal, ploy by Morgan and Adams representing the frustrating, endless, up and down nature of this battle. Westinghouse now realized that the battle was not with Edison, General Electric or Cataract, but with the House of Morgan. This was a test of wills between Morgan and Westinghouse. Even General Electric represented only a pawn on Morgan's chessboard. Morgan's focus remained on an electrical trust, not in technical or engineering victories. Westinghouse was coming to learn what so many, like Edison and Vanderbilt, had learned years earlier. Morgan's forte was strategy, not people. He looked at people as players and pawns, and to a large degree never became personally involved. He could quickly embrace an enemy for the advancement of an industrial trust. He least understood men like Westinghouse and Tesla, who mixed personal and ethical convictions with business. This lack of understanding slowed Morgan in the building of an electrical trust. Morgan held to nothing that would interfere with strategy. In fact, Morgan associates who developed strong political, religious, technical, or personal convictions became useless to him. Morgan considered passion for anything other than the end goal a liability. Flexibility in pursuit of the goal had always been a key strength in his strategies, and had perplexed Vanderbilt, Villard, Edison, and now Westinghouse. The strategy emulated out of Morgan's library, and he was a general of impressive resources. It had only been two days earlier (May 8) the Pittsburgh District Attorney had announced a grand jury for a local spying and conspiracy case against General Electric. Adams and Morgan were very confidant that they had finally end run Westinghouse. Westinghouse, however, was more frustrated than worried. Morgan's persistence and flexibility matched only his own. By ruling out both Westinghouse and General Electric it was hoped that criticism would be subdued.

Lord Kelvin would be one of the first to denounce Forbes and Cataract for unprofessional behavior. Cries continued from the whole commission and the electrical engineering profession. Adams and Sellers, however, moved forward, only conceding that he would allow new plans for the generators by all to be submitted. Adams stood strong in hopes of bringing in Tesla. Tesla may have been a quirky scientist, but he was as ethical as Westinghouse, and he would not sell out. As the World Fair progressed into the summer of 1893, Westinghouse, General Electric, and the Swiss firm of Brown, Boveri, & Company were still in the running, but only for the generators, while Forbes was to have the Dynamo contract. Pressure was mounting from legal efforts and the general outrage of electrical professionals. Tesla had also flatly refused to help Professor Forbes in the building of any dynamos. In August, while still at the Fair, Westinghouse had dispatched Louis Stillwell to discuss the overall situation. Westinghouse still held the trump cards in that his generators would probably be needed to avoid further legal complications and investor concerns.

The investor concerns usually a small matter for the powerful Morgan, took on a bigger role as the Panic of 1893 was nearing peak in August. In addition, Morgan was suffering from a personal depression with the death of his partner Anthony Drexel. Morgan was said to have spent weeks on his yacht, out of touch often with his agents. Morgan was "dazed and stunned" at the loss. Morgan had his hands full as well with the plummeting stock market that had lost 40% since the start of the year. The American treasury's gold reserve was before the mandated level, forcing the government to borrow from other foreign governments. Without a Federal Reserve, the American government needed the help of big bankers like Morgan to pull money as the United States approached bankruptcy. Grover Cleveland was secretly dealing with Morgan to put together a pool of banks by August. Morgan would have to pool with the only other power remaining, the House of August Belmont (Westinghouse's banker). In all, 158 national banks failed, 177 private banks failed, 47 savings & loan failed, and 177 state banks suspended operations, representing America's greatest depression to that point.[69] The new "Morgan-Belmont syndicate" would ultimately spare the United States bankruptcy, and start a new friendship that in 1907 would help trap Westinghouse. Belmont was a director of the Cataract as well, and Morgan needed Belmont's money to hold the project together. At least one biographer of Morgan, George Wheeler, believed that Morgan was merely a front man for Belmont and the Rothschilds.[70] Certainly, we can be assured that both Belmont and Morgan were involved in ending the panic. Morgan would gain the title of *Pierpontifex Maximus* for his diversion from trust building to save his nation's treasury.

Providence again had come to the aid of Westinghouse, but thanks to Tesla, Westinghouse had a solid technical edge as well. Westinghouse stock was falling in the midst of this panic. General Electric was down an equivalent amount and even his own firm of Drexel, Morgan &Co. was on its way to a $1.1 million dollar loss.[71] Banks failures were at record levels as well. Moving ahead with the Forbes dynamos and the General Electric AC power generators could result in a legal disaster. Morgan's lawyers were not organizational yes men, assuring him an unbiased look at the situation. The general consensus was that Tesla and Westinghouse's patents could not be successfully infringed on in the long run. Morgan surrogates had given an exhaustive effort, but the end success was doubtful. Morgan requested Adams and Sellers to work with Westinghouse on some form of compromise, at worst the strategic ploys to date had forced price concessions from Westinghouse. Forbes, however, was blinded by ego and pride thinking that Morgan was solidly behind him. Forbes started to stonewall the negotiations in August, but Adams pushed them along into September.

Adams realized that Westinghouse's technological advantage could not be beaten. Westinghouse's overall generator design would have to mesh with these new dynamos of Forbes. Forbes tried to claim uniqueness in his 16 1/2-cycle design, which was non-standard. By September, Westinghouse had a team of engineers headed by Louis Stillwell working with Cataract Construction to work out a technical agreement. Forbes was condescending and uncompromising. He belittled American engineering and Americans,

69 John K. Winkler, *Morgan the Magnificent* (Babson Park: Spear & Staff, 1950)
70 George Wheeler, Pierpont Morgan and Friends: The Anatomy of a Myth (Englewood Cliffs: Prentice-Hall, 1973), 17
71 Strouse, 334

making a statement by living in Canada. Forbes's choice of 16 1/2 cycle was efficient far from the 60-cycle current that had become standard, but the real point was that it would make Westinghouse's AC motors and rotary converters ineffective. Westinghouse had trouble figuring out if this was a ploy by Morgan to again bring General Electric in to bidding, a product of Forbes arrogance, or a demonstration of Forbes's incompetence. In any case Westinghouse was adapting having put his best electrical engineer Benjamin Lamme on designing lower cycle devices. Westinghouse was not going to be outmaneuvered by taking a hard technical stand for 60 cycles. After all, they had to adjust years earlier to Tesla's demands for 60 cycles. The best frequency Lamme and Stillwell estimated they could use to work things like the rotary converters were 25 cycles. Whatever Morgan and Adams original idea was, they were not interested in a timely and costly technical argument.

In October 1893, the Niagara project actually faced the probability of an operating failure. Adams called for a formal dinner to bring together these hardheaded scientists, industrialists, and bankers. The dinner was at the exclusive Union League Club, but prior to the dinner a lot of behind the scenes work was needed. Forbes had gone too far, but he would not compromise. Westinghouse, working with his engineers, felt that 25 cycles would work but with system inefficiencies. Lewis Stillwell, Oliver Shallenberger, Albert Schmid, Benjamin Lamme, and Charles Scott begged Westinghouse to hold out for 30 cycles (ultimately the US standard for low frequency would be 30 cycle with 60 cycle for high frequencies). Forbes remained the problem, but Adams directed by Morgan told Forbes to settle for 25 cycles. Forbes did not realize that Morgan would not have another scientist like Edison controlling and blocking progress. By October 27 as the World Fair was closing in Chicago, Westinghouse had the first part of the contract but Forbes continued to be a problem.

Westinghouse had the contract for three 5000 horsepower dynamos, which had to meet the specifications of AC current, two-phase, 25 cycles and 2200 volts. Forbes, however, remained upset with the compromise of 25 cycles. He belittled the Westinghouse engineers whenever possible. In addition, George Westinghouse believed Forbes was moving towards creating his own dynamo to compete with Westinghouse. Westinghouse engineers were uncomfortable sharing details with Forbes. To accomplish anything Dr. Coleman Sellers started to function as the bridge between Adams and the Westinghouse engineers. Sellers, of course, was an Adams consultant, but he also was highly respected by George Westinghouse and his engineers. Finally, in December 1893, Westinghouse had had enough telling Cataract; Westinghouse Electric could no longer work with Forbes. Adams and Morgan then decided to cut Forbes out of the decision making process. Forbes was a puppet and did not have the leverage of Edison, and Morgan would not jeopardize the project (or syndicate profits) over some technical issue. Technical arbitration would be handled directly by Coleman Sellers. Forbes now knew that he was only a pawn in the struggle, not the king he had thought.

Later he went to the press calling the American engineers ignoramuses, and Cataract management incompetent. Westinghouse could finally move forward.

Although Westinghouse got the largest part of the contract, General Electric also ended up with some lucrative contracts for transformers and transmission lines. General Electric work on the transformers became critical to Niagara's success. GE had a slight operating advantage in the construction of transformers. First, they owned Professor Thomson's oil immersed transformer patent of 1887. In 1893 GE engineers made further improvement to the insulation that "constituted a significant step in transformer development."[72] These transformers stepped up voltage from a generated 2000 volts to 10,000 as well as change the two-phase current to three phases. The current at 10,000 volts was then transmitted through bare copper cable (on thousands of wooden insulated poles) to buffalo, twenty-six miles away. GE transformers then stepped down the voltage to 110 volts for general lighting and 370 volts for Buffalo's street-railway system. In the end Westinghouse's great victory generated vital profits for General Electric as the Panic of 1893 continued to 1896. With Sellers, General Electric and Westinghouse started to work together for the good of the Niagara. Morgan as well realized any further internal competition would cost the overall project and his syndicate money. Morgan knew the war over Niagara was over, and it was time to move on.

Westinghouse won the war because he was a better engineer and manager, who built a superior organization. Edison's stand on DC current had cost General Electric too many loses in the long run. Edison was a brilliant scientist, but could not manage other scientists and engineers, and worst, he would not listen to them. In addition, Edison had a primitive approach to the complexity of Victorian electrical engineering. Edison saw little need for theory, preferring experimentation, which worked well for chemistry, physics, and biology, but bogged down as the scientific complexity increased. Tesla explained the difference between Edison and Westinghouse's approach.

> If Edison had a needle to find in a haystack he would proceed at once with the diligence of a bee to examine straw after straw until he found it. I was a sorry witness of such doings . . . a little theory . . . would have saved him ninety percent of the labor.[73]

Even worse, Edison would have five expert scientists stand behind him as he handed them each straw so they could keep count. Tesla represented a new breed of engineer, which required a new breed of engineering manager, such as Westinghouse. Actually, J. P. Morgan was one of the earliest to realize the new breed.

As soon as the Niagara contracts were let, Morgan and Adams again looked at other ways to improve General Electric's position in the overall war. Adams was sent to talk to the man that had caused their defeat. Nikola

72 Hammond, 238

73 M. Cheney, *Tesla: Man Out of Time* (New York: bantam Doubleday Dell, 1981)

Tesla had been continuing his work on a variety of electrical devices in his New York lab. Tesla felt free now to deal with these new patents with anybody. In fact, Westinghouse did not have the money available to buy any more patents. With Westinghouse's hands and money tied, Morgan and Adams wanted to bring Tesla into the fold. Tesla needed money as well, not so much for his lab but his high living. Tesla had a number of new patents for wireless transmission, refrigeration, fluorescent lighting, and remote control. Adams offered Tesla $100,000 for Tesla's new patents and controlling interest of a new company for future patents. In February 1895 Tesla Electric was formed with Edward Adams, Adams's son, and William Rankine as investors and directors. Tesla expanded his New York lab, but had difficulty getting some equipment from Westinghouse Electric. Tesla would be a life long friend of George, but he had many old enemies in Westinghouse Electric, such as Albert Schmid. In the long run most of Tesla's commercial inventions were behind him, but Morgan was taking no chances that some future Westinghouse-Tesla partnership would blindside him.

1895 started out as a different year with the death of George's mother at Solitude. She had been living with George, Marguerite, and young George III for five years. Westinghouse as usual dealt with setbacks by throwing himself into his work, and 1895 represented one of his busiest. He was in the process of building the equipment for Niagara and moving Westinghouse Electric into its new plant in East Pittsburgh. Westinghouse was expanding his plants in England, Russia, and France. For 1896 new plants were planned for Austria, Belgium, Balkan Islands, and Germany. For example, he had over 7,000 employees in England alone. Westinghouse Machine was expanding into large steam engines and was planning to build a new production plant in Canada. None of the great industrialists of the time like Carnegie, Frick, Schwab, and Vanderbilt managed the size and diversity represented by the many companies of Westinghouse. It seems mind-boggling, the diversity as well as the geographical spread. Westinghouse often seemed to be living out of his "Pullman Palace," as well as in steamers across the ocean. 1895 can however, be considered his peak year or at least the year of his greatest triumphs. Many wanted the great man to speak, but his fear of speaking kept this to necessary award banquets.

The first power came on at Niagara on August 26, 1895. On that August day "Dynamo No. 2" came online supplying current to the Pittsburgh Reduction Company. The initial power went to the electrochemical plants nearby. Dynamo No. 1 didn't come on stream until September 30, at which time the Directors of Cataract and Westinghouse plus assorted millionaires came to celebrate. Two important figures were not present. One missing was J. P. Morgan who came later in October, and the other was Nikola Tesla. Westinghouse, as the plaque at the Falls would reflect, gave all the credit to Tesla. The *New York Times* found a way to hail the event without mentioning Tesla, while most other papers hailed Tesla as the equal of Edison with the triumph of the Falls. It was a great victory, but Westinghouse spent time at Pittsburgh Reduction further studying the wonders of aluminum smelting versus rather than celebrating.

One of the great results of the Niagara Power station was the rise of the American electrochemical industry. The two biggest customers of Niagara were the Pittsburgh Reduction Company and the Carborundum Company. Pittsburgh Reduction Company (the future ALCOA aluminum) contracted for 5000 horsepower of the 15,000 available. Within a year a massive electrochemical industry grew up around Niagara including sodium, soda ash, sodium peroxide, and calcium carbide plants followed. By 1897, 12,500 horsepower of the 13,500 horsepower available for local consumption was for electrochemical plants.[74] Westinghouse enjoyed visiting these plants, since he always had a fascination with metallurgy. The aluminum industry owes a great deal to George Westinghouse.

Aluminum requires huge amounts of chemical or electrical energy to unlock the aluminum from its ore. The metal had caught the imagination of Victorians during the 1855 Paris Exhibition. A few button size pieces, which represented the world's total inventory, launched a quest to apply this unique metal. The French metallurgist Sainte-Claire Deville had developed a complex chemical process to produce these quantities. Deville had brought the price down from $540 a pound to about $16 a pound. These simple pieces at the Paris Exhibition fired the imagination of the likes of Jules Verne, Charles Dickens, and Napoleon III. Napoleon III, a lover of new technology, ordered aluminum tableware for his best quests to replace silver and goldware. He gave Deville the money to build a production plant to outfit the French army in aluminum armor. Commercial development was just too expensive; in 1882 the world's production was just a few hundred pounds, yet interest and demand grew. A young Ferdinand von Zeppelin could not find a supplier to manufacture an aluminum dirigible because of the limited production available.

The breakthrough came in 1886, when Oberlin College student Charles Hall developed an efficient electrochemical process. One drawback to aluminum smelting with the Hall process, even today, remains its consumption of electrical power. The amount of energy needed to smelt just one pound of aluminum would keep a 40-watt light bulb burning for 10 to 12 days. To state it another way, today aluminum smelting takes 4% of the world's electrical power. In 1888 Charles Hall teamed up with Pittsburgh metallurgist to form Pittsburgh Reduction Company. On Thanksgiving Day, 1888, two large Westinghouse dynamos driven by a 125 horsepower Westinghouse engine started the production of aluminum at about $3 a pound. This plant helped Westinghouse see the future of electricity for aluminum smelting. It also helped Westinghouse change his mind on the use of compressed air at Niagara Falls. The success of Pittsburgh Reduction made it clear that there would be a major customer awaiting Niagara electricity. Part of Westinghouse's

74 Passer, 293

vision for the future of electrochemical aluminum smelting came from his belief in the future of the metal itself.

Westinghouse anticipated that cheap power would mean low cost aluminum, and he and his engineers were looking at potential uses. Aluminum additions to Babbitt bearing metal showed superior results at Westinghouse Machine Company. These engineers were also using aluminum bronze in corrosion applications. Westinghouse soon became a customer of Pittsburgh Reduction experimenting with aluminum parts. One of his first applications was a valve in his air brake. He also found uses of it as a shim for machinery. Westinghouse's foundry even looked to it as a casting metal to replace cast iron. Another application that Pittsburgh Reduction Company was the first to use aluminum foil as a wall decoration. Westinghouse had studied the production of aluminum foil in France, and had suggested to Pittsburgh Reduction its manufacture. Westinghouse and a number of his Homewood neighbors, like Henry Clay Frick, used aluminum foil on walls. The Frick mansion still has the aluminum wall decoration of the period observable today. Aluminum was unusual in that creative people, such as Jules Verne had already shown its future uses. It was a metal with demand awaiting a production means. Pittsburgh Reduction would be one of the first to sign on for Niagara's "electrochemical city." The low cost of Niagara aluminum opened up new uses in art. In 1898 famous British artist Sir Alfred Gilbert cast several largest cast aluminum statues of St. George (the patron saint of England). The biggest use of aluminum remained the steel industry were it was a key additive to cast steel, another application Westinghouse had studied himself in the development of the steel railroad frog.

Another big customer of Niagara power was the Carborundum Company. Carborundum was an artificial abrasive of aluminum oxide. These alumina crystals were produced by electrochemical processing like pure aluminum. Larger colored alumina crystals in nature are sapphires and rubies. These fine artificial alumina crystals are extremely hard, making them ideal for grinding and cutting wheels. Within a few years Niagara had become a major production area for steelmaking ferroalloys, such as ferrosilicon and ferrochrome. The massive amount of workers needed to build the power system and operate the electrochemical plants required a planned worker town of houses, which would be the first in the country to be lighted and heated by electricity. Industrialist William Love hoped to emulate Westinghouse's Wilmerding with an industrial utopia for electrochemical workers. The effort resulted in a mile long town along the wastewater canal known as "Love Canal." The town failed to grow and became the LaSalle area of Niagara. It would become famous for the old toxic electro chemicals that caused a strange "outbreak" of cancer in the 1970s. While Niagara first powered this great new industry, it would not be until November 15, 1896 that power would flow the twenty-six miles to Buffalo, New York. The first customer in Buffalo was the streetcar company, and it was streetcars where Westinghouse's interest had started to be directed.

Chapter 13. East Pittsburgh

The Niagara project had been one of the greatest competitive battles in the annals of American business. Westinghouse's triumph came from a combination of innovation, respect for intellectual property, a superior technical team as well as a great legal team, and with a bit of luck. In November of 1895, Westinghouse would start an eighteen-month effort to deliver the three behemoth generators to Niagara. Westinghouse would be personally involved in the daily work to complete the project. The potential for electrical generation was exponential in 1894. For every thirty-five homes illuminated by gas, the market had only one lit by electricity. One of George's personal projects of the period was to convert his neighbor Henry Clay Frick's home from gas to electric light. The last five years of the 19th century saw a new burst of creativity that went in new directions. He would sign a peace treaty with General Electric that divided up the market until its was declared legal in 1911. Westinghouse Machine Company would build ever-larger steam engines, and pioneer other fuels such as natural gas and fuel gas engines. Westinghouse Air Brake would advance new workers programs, but his passion would be in his Westinghouse Electric company projects. These would include launching into the area of electric traction with the New York Interborough Rapid transit and Chicago's South Side Elevated Railroad. This would require him to expand in the Turtle Creek valley with his huge East Pittsburgh plant. He would again use the rudiments of his childhood dream of a rotary engine to develop the first turbo-generator. He would develop some of the first fluorescent lights. Some have noted that his productivity in the number of patents issued dropped off during this period, but what is overlooked is the diversity of his effort.

His immediate concern in 1894 was the completion of the mammoth generators for Niagara. The production of the Niagara generators strained the Allegheny Works (main plant on Pittsburgh's North side) and Garrison Alley (often referred to as the Pittsburgh Works) as well as the organization of Westinghouse Electric. Westinghouse realized that as he moved into power generation and his new pet project electric traction, he would require a much larger plant and organization. He purchased 40 acres in East Pittsburgh for a future plant site, but early in 1894 he started the development of a new organization. The growth of Westinghouse Electric and the intensity of the battle for the Chicago World Fair and Niagara had allowed a patchwork organization to evolve. Departments and teams were organized to solve specific problems or develop needed devices. Research had become uncoordinated and there was no engineering department. The Garrison Alley had regressed to the uncoordinated style of Edison's old Menlo Park. Westinghouse needed technical managers and new young engineers to develop a true engineering department. Managers, like P. M. Lincoln from Brush Company, and young engineers, like Bertha Lamme (America's first women electrical engineers).

The generator production at the Allegheny Works (Pittsburgh's North Side) and Garrison Alley created a new set of headaches for Westinghouse. Aisles were far too small in the factory and had to be widened. Cranes had to be upgraded to lift the weights demanded by the generator parts. Westinghouse like his work at Westinghouse Glass to grind stopper lamps showed brilliance. Westinghouse and his engineers took the original generator design, which the factory lacked the capability to make, and redesigned it for manufacture. Today many talk about Chrysler Corporation's DFM (Design for Manufacture) as revolutionary, but Westinghouse used the technique almost a hundred years earlier! Jill Jonnes noted, "The original intended size and scale of the generators had to be reduced to ensure that they could be hoisted onto a railroad flatcar and transported safely to Niagara."[75] As Westinghouse adjusted generator design to meet factory capability, he was designing his new East Pittsburgh Works. The aisle of his new factory would be the world's largest being 70 feet wide and a third of a mile long. The cranes would also be the world's largest with fifty-ton lift capability and the ability to move the full length of the manufacturing aisle. The Pittsburgh Press hailed it as "the largest and most modern workshop in the world."

Westinghouse pushed hard to complete his new East Pittsburgh plant by the end of 1894. The plant was a huge undertaking and required huge sums of capital, which he depended heavily on from his banking syndicate. Pittsburgh and the nation were in a great recession, but Westinghouse continued to forge ahead. In the summer of 1894, Pittsburgh was to host the an-

75 Jill Jonnes, 306

nual national convention of the Grand Army of the Republic, an organization that Westinghouse was active in. he had often had local meetings at Solitude using the extensive grounds for picnics. The city of Pittsburgh needed a meeting location for 6500 plus attendees. Westinghouse volunteered his East Pittsburgh partially completed plant. Using his own money he build a stage and huge dining hall. He paid completely from his own funds to supply meals for several days for the Civil War veterans. This was the first use of the plant before it was pressed into operation to build the needs of Niagara.

Besides East Pittsburgh Works being a manufacturing plant for such things as electric motors, it was also the largest plant in the world to use electric motors. Westinghouse built his own power plant for east Pittsburgh to produce AC current for general-purpose electric motors throughout the plant. Surprisingly, the East Pittsburgh power station generated 25 cycle current not the now standard 60 cycle. The reason was that the slower speed 25 cycle AC motors had the higher torque needed in applications such as crane motors. Westinghouse had first used the higher torque 25-cycle motors for cranes and hoists at Westinghouse Machine. Of course, 25-cycle frequency equipment was the result of the Niagara compromise with Forbes. Electric motors were used for the first time to drive metal machining equipment, eliminating the leather belt/steam arrangements needed in machine shops. The communications system of 600 phones and compressed air tubes to send blueprints was state of the art as well. East Pittsburgh represented the move of American factories from the steam age to the electrical age. While the overall dimensions impressed the media, the application of electric motors was revolutionary. At the International Institute of Electrical Engineers a 100-page guidebook was published in 1904 for international visitors to the East Pittsburgh plant, which highlighted the many wonders for the engineers. The East Pittsburgh's power plant would also produce DC current via rotary converters to run an electrical train for the plant. The success of heavy duty electric motors at East Pittsburgh would lead to a new national standard of 60 cycle for lighting and small motors, and 25 cycle for heavy duty applications in industry. One major but over looked improvement by Westinghouse in AC motors was the use of ball bearings to reduce rotating friction in 1901. The study and improvement of ball bearings had been one of his many hobby-like pursuits over the years. His improvements in bearings further increased the efficiency of Westinghouse motors over the competition. These bearing improvements allowed for extremely quite motors that could be applied in household appliances. It would not be until the 1930s that the competition caught up. By 1920, with further motor improvements, the standard for all motors would once again become 60 cycles per second for all applications.

East Pittsburgh had 5,000 employees in 1895 and 6000 employees by 1897. The factory design took the individual worker into account as he had in his Wilmerding Air Brake plant. Washrooms and toilets were designed with the same creativity as electrical devices. Dining rooms were built and Westinghouse subsidized meals to assure reasonable rates. The plant was the precursor of ergonomic design with skylights to maximize daylight, which was estimated at 70% of outdoors. Bowling alleys and poolrooms were constructed. There was an auditorium for movies, shows, and lectures. A library was added as well with popular books and magazines. Overall lighting was the best available including vapor lamps. There were a number of conference rooms for the many employee clubs and organizations. One employee club, The Electric Club, would become world famous rivaling the greatest technical societies of the time. The club offered two lectures a week on technical and popular topics, as well as printing a monthly newsletter. The newsletter quickly evolved into a technical journal of international distinction. Westinghouse often allowed outside organizations to use the conference rooms. In 1902 a large assembly hall was built for the Electric Club in the near-by suburb of Wilkinsburg. The Electric Club was open to all Westinghouse company employees and even included some non-Westinghouse area engineers. Every two weeks the Electric club had a social event planned from a committee of members' wives. One Christian group offered a Thursday Bible study, another Christian group, Christian Endeavor, was also given access to the rooms. Christian Endeavor was a national evangelical movement of the period.

Westinghouse believed in fraternity and community as foundational to productivity, and East Pittsburgh had other clubs related to the plant. There was a Foremen's Association, which had 200 to 300 middle managers. It included foremen, Chief clerks, and inspectors. The association was both social and professional as was characteristic of Westinghouse organizations. The Casino Building erected near the plant offered recreational and educational opportunities for East Pittsburgh employees. The Casino functioned as a corporation school or night college offering general courses in English, shop practice, mathematics, and science. Westinghouse Electric funded 70% of the budget, but students were charged $2 per month. Here again we see a fundamental belief of Westinghouse about charity that employees should pay for education. The courses were offered on a two-semester basis and averaged over 200 employees. The Casino building included bowling alleys and billiard tables. Another organization was the "East Pittsburgh Club" designed for comfortable dining for the engineering and sales staff. The social and dining elements were things Westinghouse felt important as described in the following:

> On the theory that close acquaintance leads to better understanding and more cordial co-operation, the company maintains two beautifully furnished dining rooms on the sixth floor of the Office Building where the executive officers, managers and representatives of each department daily gather at the noon hour for a substantial lunch and a free interchange of ideas all the more effective because it is unofficial. These dining rooms have

> done much to promote the harmony and kindly spirit of helpfulness so characteristic of this organization. [76]

This practice was so successful that much of Pittsburgh's heavy industry adopted these management approaches.

East Pittsburgh and Wilmerding earned the interest of many traveling criminals of the time as well with the large Westinghouse paydays. A group of "tramps," "con-men," and "streetmen" rode the rails to prey on workers. These traveling streetmen set up gambling wheels and games by the plant on paydays. Four plants were rated as the "best works in the country in which to make a pitch." These works had large payrolls and were located on the major railroad junctions. They included "The Baldwin Locomotive Works in Philadelphia, the Union Iron Works in San Francisco, the Pullman Car Shops in Chicago, and the Westinghouse plants of East Pittsburgh."[77] Wilmerding was able to control the problem, but the massive East Pittsburgh complex and local police could not. Like a carnival, the games were "legal." The traveling con men targeted the big holidays in many cases. In fact, other than angry wives, the immigrant workers seemed to enjoy the gambling. The prosperity of the plant seemed to foster an acceptance. East Pittsburgh was, in fact, one of America's highest paying plants for piecework, and a very successful one as well.

The procedures, methods, and management approaches were just as progressive at East Pittsburgh. The quality control practices employed were far ahead of their time, but evolved out of Westinghouse's desire to maintain the Westinghouse image of quality and reliability. All supplied materials were subjected to extensive testing. Steel and iron not meeting standards was routinely rejected. Steel chemistry specifications were often set by Westinghouse engineers, as well as metal formulas for things like babbitt and bronze. Every order of motors was subjected to a random inspection, which could result in the full rejection of lot if standards were not met. Testing records were available to the customer, and the company welcomed customer on site inspections and auditing. Many of their top of the line motors received a 100% inspection. These tests included operating horsepower measurements, efficiency measurements, and appearance items. Westinghouse even did joint testing and in service testing of the motors with companies like Baldwin Locomotive.

Westinghouse seemed to be besting the Morgan trust on all fronts. The strength of Westinghouse Electric had come from Westinghouse's concept of organization, which has been overlooked because of Morgan's negative publicity campaign. Westinghouse was the conductor of a technical orchestra. Westinghouse developed men to power his technical conquests. He always showed humility in his corporate successes crediting all involved. The simple plaque at Niagara Falls hung in 1899 credited thirteen patents, none

76 Pittsburgh Electrical Handbook (St. Louis: American Institute of Electrical Engineers, September, 1904), 74

77 Wage Earning Pittsburgh, Survey, 1914

of which were his. He credited 9 patents of Nikola Tesla, his associate and friend; several of his research directors, Albert Schmid; and two patents of his competitors. Sometimes simple statements tell us more than volumes of testimonials. The plaque listed the names and patents numbers that made the great engineering feat possible under the name of Westinghouse Electric. This reflected a great respect for individual contributions and a respect for intellectual property rights. Westinghouse focused on developing his organizations one person at a time. His managers were selected for their ability to train as much as their ability to lead. His plant managers often spent considerable time lecturing and training on technical subjects. A department manager was expected to develop his employees in similar manner. This was particularly true at east Pittsburgh where employees dealt with the full complexities of electrical engineering.

Westinghouse's research manager Benjamin Lamme described his experience as a department head at East Pittsburgh:

> After I entered the Westinghouse Company, I was more or less forced into the educational field, through the nature of my work. I had to build up methods of analysis and calculation; and, as the work grew, I found it necessary to train others in such matters. In order to obtain the greatest amount of assistance from others, it always appeared to me necessary to educate them in the work just as far as they could go. In consequence, although the process was slow through the early years of my connection with the company, I gradually built up a force of young men to whom I endeavored to impart practically all of my own experiences and methods. Also, as I have said before, I never adopted the policy, too often followed, of declining to have around me men whom I thought might possibly go ahead of me. I have always felt that the stronger I made my particular department, the stronger I, myself, would be; and after all of these years, I have found no reason for changing my notions on this point.

Lamme represented the model manager of Westinghouse's vision, part trainer and part leader. Even more importantly, Westinghouse managers were expected to hire the best. Managers were promoted as much on the performance of their department as their personal performance. Westinghouse pioneered in his view of a technical manager. He demanded that a manager take as much pride in the accomplishments of the employees as his own. A manager was to impart every piece of this knowledge to the employees. Westinghouse had this skill, but as he found out few others did. Still, the pure engineer, such as Tesla, always had a place too in his organization.

The roots of this educational approach to employment are deeply imbedded in Westinghouse's own experience as an apprentice in his father's machine shop. Electricity offered the same complexity as machining, and Westinghouse knew the value of one on one training. He also believed in achieving skill levels in a progressive manner. Electricity offered a greater challenge in its theoretical base. To accommodate this Westinghouse split his apprentice program into a practical electrician program and engineering program. The training was free, but required a significant time commitment from the apprentices. The training was very personal with department heads and experts teaching. The East Pittsburgh plant often took on the character of a community college, and soon became a national resource for electricians, engineers, and technical managers. Westinghouse men dominated America's technical societies and journals bringing East Pittsburgh international fame. A Westinghouse training certificate was highly respected throughout the world. East Pittsburgh formed the historical model for a learning organization.

Between 1895 and 1904, East Pittsburgh became a massive Westinghouse manufacturing complex, including Westinghouse Electric, Westinghouse Machine Company, and the near-by Trafford City Foundry (Trafford was three miles away) that supplied both great organizations. This part of the Turtle Creek valley was the engineering Mecca of the world, with almost hundreds of engineers visiting daily. No greater manufacturing valley has ever existed for heavy industry. In its peak "Westinghouse Valley" would employ over 30,000 employees. The valley boasted world-class foundries of iron, steel, and brass, as well as electrical equipment factories. The automation was the best in the world, foreshadowing the great assembly lines of Ford. The management system was one of the first in the country to fully implement the scientific management concepts of Fredrick Taylor. Westinghouse was correct when he questioned the critics of the press in 1907 as he passed thorough on the train to turn Westinghouse Electric over to bankers. This manufacturing complex of the Turtle Creek Valley rivaled anything in the world. Certainly, it deserved to be one of the seven wonders of the industrial world, which would include the Rhine Valley of Krupp in Germany and the Monongahela Valley of Carnegie.

Besides the great generators of the era, Westinghouse East Pittsburgh produced more electric motors than anywhere in the world. The use of those motors was diversified, but the original interest of Westinghouse was eclectic traction, or streetcars as they came to be known. Electric traction was a natural extension of his dual interests in railroads and electricity. Westinghouse was not the first in the field, but once he applied himself, the company would become a major factor as was usual of Westinghouse goals. Westinghouse had seen the earliest endeavors into electric traction on his many trips to Europe. The famous German electrical inventor Werner Siemens, had pioneered electric traction with a half-mile electric railroad at the Berlin Exhibition of 1879, which Westinghouse had ridden and inspected extensively. In 1881 Siemens built a six-mile line in Ireland. America lagged behind Europe; its first track came in 1887 with Bentley and Knight putting in a short line in Allegheny City (Pittsburgh's North side) near the Westinghouse Electric plant! This company became the Observatory Hill Passenger Railway and covered four miles. The company was said to have gone bankrupt over the cost of copper wire and brushes needed. The copper brushes were costing over ten dollars a day alone. Westinghouse enjoyed riding the Bentley and Knight when at the plant, which contributed to his growing interest. Westinghouse often observed the need for horses to pull the car up the steep

Observatory Hill. Westinghouse was also quick to observe the build up of copper on the rails from the contract brushes. In fact, brushes were being consumed daily at a rate too fast for the production plant to keep up. Frank Sprague put a 12-mile, 40-car system in 1888 at Richmond, Virginia.

Frank Sprague was a graduate of the United States Naval Academy in electrical engineering, who had actually joined Thomas Edison. Sprague as a graduate engineer saw little opportunity in Edison's autocratic technical camp. Sprague reflected Edison's inventive diversity inventing over sixty devices, such as an improved telephone, machines to send pictures by wire, and numerous telegraphic devices. In 1884 Sprague broke with Edison to do development work on DC motors. Sprague, however, needed to subcontract Edison for his manufacture. In 1884 he formed the Sprague Electric Railway and Motor Company, which pioneered the use of electric DC motors in industry. Applications included printing presses, cranes, water pumps, freight tramways, elevators, and streetcars.

Westinghouse had first seen the Sprague system at the Philadelphia Electrical Exhibition in 1884, but at the time, it was merely a motor with a promise of a future. The Sprague motor found many applications by 1889 and manufacturing was subcontracted to Edison. Sprague continued to push their use in electric traction with good success. By 1889 there were 180 streetcar railways and 1260 miles of track.[78] In the fall of 1889, Westinghouse assigned Benjamin Lamme to explore the streetcar industry and prepare a report for him. Basically Sprague was functioning as a distributor of an Edison manufactured motor for streetcars. Streetcars were changing the nature of urban centers allowing larger growth in the suburbs. New York city posed the biggest resistance to streetcars fearing the traffic congestion as well as the sparks and potential dangers. The congestion concerns would ultimately lead to elevated systems in big cities. With the formation of Edison General Electric in1889, Sprague Company was merged with Edison General Electric. The new president of Edison General Electric, Henry Villard, didn't see the growth potential of the Sprague streetcar system believing a safer lower voltage system would be needed and the consumption of copper brushes would have to be addressed. It would be one of the few times a Morgan man would so miss the market potential. In 1888 there were 130 cars in operation, but in 1892 there were 8000 (there were still more horse streetcars). In ten more years in 1902, there would be 60,000 cars! The market had many interested besides Sprague. Sprague resigned over an agreement with Villard to call his system, the Edison System. At the same time, Thomas Edison had no interest in or time for the development of a streetcar system.[79] By 1890 Edison General Electric was out of the streetcar business (it would come back in 1894 with the merger of Thomson-Houston).

78 *Electrical World*, October 19, 1889, pp 264-265

79 F. Sprague, "Digging in the Mines of the Motors," *Journal of the American Institute of Electrical Engineers*, May, 1934

Charles Coffin, then president of Thomson-Houston Company, saw things much differently. Coffin believed that possibilities for beating Edison General Electric in incandescent lighting had disappeared and electric traction was an area for Thomson-Houston to exploit. In 1888 Thomson-Houston began to explore the possibilities. Charles Coffin observed the same things that Westinghouse had. Westinghouse had already talked of the need for a bigger 15 horsepower motor for steep grades. Both Coffin and Westinghouse realized that the copper brush usage problem would have to be resolved. Coffin insights were surprising, since his enemies pointed out his lack of technical background by calling him "the shoe salesman," which he had been prior to coming to Thomson-Houston. Maybe not so surprising was the brilliant scientist Professor Thomson working for him. Coffin, however, opposed Thomson plan for an internal development program and purchased the system of Charles Van Depoele. Coffin, Thomson, and Westinghouse had all ridden Depoele's system at the Chicago Industrial Exhibition of 1883. The Depoele system had a beefy 30 horsepower motor and superior mechanical devices for the streetcars. By 1888 Depoele had systems operating in Detroit, Michigan; Montgomery, Alabama; Dayton, Ohio; Harrisburg, Pennsylvania; Scranton, Pennsylvania and ten other cities. In purchasing Depoele's system, Thomson-Houston had a proven system.

Still the copper brush usage had to be addressed, but Depoele would resolve this as well. Depoele was brought in as consulting engineer and would receive a royalty. Depoele started experiments, which led to the use of carbon brushes to replace the copper. The improvement would eliminate sparks as well as the loss of copper. While Thomson-Houston owned the patent on carbon brushes, they allowed the smaller competition to use it. The policy established is what was known as free usage, making later patent infringement lawsuits difficult. Even when General Electric brought Thomson-Electric in 1892, they did not pursue the patent rights, allowing Westinghouse to take advantage of the technology.

Generally these electric trains, trolleys, or streetcars operated on 500 volts of direct current. The original term was electric traction or train, but was quickly replaced with the term trolley, referring to the contact device that brought current from the overhead wire. The "trolley" was a two-grooved wheel device that was pulled along the wire. A grooved wheel on the end of a pole soon replaced the trolley device, but the term trolley struck. Generally the technology consisted of a high voltage usually alternating current (over a mile) of 2000 volts from the generating plant to a substation. At the substation the alternating current was changed to direct current and stepped down to 500 volts. The direct current was then distributed by a "trolley" wire overhead and steel rails. The trolley or streetcar received the current via the grooved contact wheel on the end of a pole and completed the circuit with copper brushes touching the rails, which operated a double reduction DC motor. The trolley operator could control the on/off switch and motor speed. Mechanical brakes were also used. In 1889 Tesla, Westinghouse, and

Schmid had tried unsuccessfully to use the Tesla AC motor directly at the Garrison Alley plant. As Westinghouse experimented, there were three major players in the field: Sprague, Thomson-Houston, and Short.

Interestingly, trolley development was advancing during the Edison publicity campaign against Westinghouse AC systems. While the trolley systems were DC from the substations current did flow through two steel rails and the overhead wire. Sparking of the contact wire and rail brushes was common. It is even said that fearful residents of Brooklyn "dodging" trolley tracks was the source of the baseball name "Brooklyn Dodgers." Westinghouse even did tests with old horses, to discover whether their hoofs touching both rails would cause a shock. The system proved safe. One of Westinghouse's oldest engineers recalled Westinghouse during the experiment:

> The very first time I entered the Garrison Alley factory in the fall of 1891, it was my privilege to see George Westinghouse in action. He wanted to eliminate trolley wires from the streets and to collect current for cars from contacts or buttons projecting slightly above the pavement between the rails. He wanted to find what voltage would be safe on the exposed contacts. A short track had been arranged and a horse was ready. He was knocked flat by 110 volts; lower voltages were nearly as bad, and even 10 volts excited him greatly. The test stands out in my memory for two reasons. First, I saw Mr. Westinghouse in action, an experience that thrilled me; and second, I have always felt that although direct current was used in that test, Mr. Westinghouse had in mind, even then, the use of single-phase alternating current with low voltage on the contacts.

While many of the streetcar devices were patent protected, the system itself had evolved with little patent protection, allowing Westinghouse to enter the field. Again the reorganization of General Electric, the battle for Chicago and Niagara, and the nation's financial concerns seemed to have distracted Morgan as Westinghouse took aim at electric traction. Actually, Westinghouse back in 1887 had hoped to develop an AC system using the Tesla motor. Tesla worked at Westinghouse's Garrison Alley lab to develop it, but they couldn't fully bring it into reality. Westinghouse then asked his lab manager Albert Schmid to assemble a team to develop their own DC system. Westinghouse set the objectives of a better motor, remembering the need for horses on hills, and improved safety, which had attracted politicians (a number of state legislatures were looking at laws to outlaw overhead trolleys). In May of 1890, Westinghouse started to market his own system that varied little from that of Sprague and the Thomson-Houston systems. The Westinghouse engineers started to develop new motor ideas motivated by their boss. Westinghouse, when in town would stop at the lab to ask about progress or drop off a new idea to try. The noise and inefficiency of the double reduction gear motors in use was an area that George felt could be improved upon. He brought in some of his best mechanical engineers from his other companies including Herman Westinghouse, using his dinner parties for cross-fertilization of the project.

Westinghouse first tried a gearless motor driving the axle as an extension of the motor shaft, but it applied too much power. Finally, a single gear motor was developed and he started to market "Westinghouse No. 3" in 1891. Benjamin Lamme was the proud developer of the No. 3. This forced Edison General Electric and Thomson-Houston into a motor development race. This race, unlike the war of the currents, was between the firms' engineers and was a matter of pride. The battle of Chicago and Niagara was Westinghouse versus the Morgan team; this was an engineer against engineers. Harold Passer described it best, "The individual incentive was supplanted by a group or team incentive such as is characteristic of athletics. The General Electric railway engineers saw the Westinghouse railway engineers as one football team sees another and worked together to try to defeat them. The judges in this contest were the street-railway managers who brought and used the motors and, through their buying, gave the decision to one engineering staff or the other." I can think of no better analogy for George Westinghouse than that of a football coach. It explains his focus, intensity, and sometimes rough comments if things weren't going right, yet his encouraging leadership when needed. After years of research, reading this football analogy, I thought, yes, that finally explains Westinghouse to me. I could understand the sometimes-paradoxical behavior, like a coach it depended on the game. Too many times the press seemed to make it personal, but for Westinghouse it rarely was personal. It was the competition. In this respect Westinghouse and his adversary Morgan were the same.

General Electric's answer to the Westinghouse No. 3 motor was the GE-800. The experimentation on motors was endless at both General Electric and Westinghouse Electric. As his engineers battled over improved motor design, Westinghouse continued to look at the overall system. He set up a test track in the vacant lot behind Solitude to test a surface conductor system to eliminate overhead trolley wires and in 1895 setup a four mile experimental track around his East Pittsburgh plant. As late as 1904 Westinghouse was still experimenting with AC motors, actually bringing the system into use in the New Haven Railroad. All this constant experimentation led to the "button" system. The button system used contact shoes to pick up current only when a button forced contract of the shoes. It was a safer system, but General Electric would challenge the patent rights, and Westinghouse would ultimately lose. The legal fight was just one of many at the time between General Electric and Westinghouse on electric motors and traction. Battles started to break out in elevated electrification projects of large cities, such as Chicago and New York. The streetcar and electric train market boomed from 1887 to 1902. In 1887 there were 1300 miles of track and 80 cars; by 1902 there were 22,000 miles of track and 51,000 cars. When Thomson-Houston merged into General Electric in 1892, the battle between the remaining giants was on. The competition was getting stronger as General Electric's organization was improving to catch up to that of Westinghouse.

Streetcars were changing the very fabric of American society. Horse drawn streetcars averaged a mere six miles per hour compared to about twelve miles per hour for the electric streetcar. In residential areas, streetcars could achieve twenty miles per hour. Streetcars started the first phase of growth for the suburbs, often extending the population growth a distance of four times further from the central business district. Even more amazing was that the cost per mile of the electric streetcar was half that of the horse drawn car.[80] The streetcar had appeared early in the 1870s in Pittsburgh as horse drawn cars. The Turtle Creek valley appropriately had one of the first electrified systems in the Braddock and Turtle Creek Railway Company. In the 1890s, streetcar bridges started to unite the families of Pittsburgh separated by swift rivers. Carnegie and Mellon supported these streetcar companies, and by the turn of the century. It was estimated that 40,000 workers were transported daily to and from work. Thanks to Carnegie, the great streetcar terminus was at Kennywood Park, and thousands rode streetcars to get there every weekend. On a Sunday afternoon the streetcars turn around could rival the country's busiest railroad station. Streetcars continued as a means of transportation throughout the Pittsburgh area into the 1980s. Kennywood Park was about fourteen miles from Pittsburgh and about three miles from Wilmerding.

Heavy electrified trains, although considered a different product than that of the streetcar, but a logical extension of product lines for both General Electric and Westinghouse. In 1895 Westinghouse moved into partnership with Baldwin Locomotive to build AC electric locomotives to compete with General Electric's DC. The competition would again be cutthroat, but it would be eliminated with the Diesel locomotives of 1910. George Westinghouse believed in cooperation and the Baldwin Locomotive partnership was an example of this. General Electric locomotives were in competition with Baldwin steam locomotives and Westinghouse Electric trains. Westinghouse, as shown, found this strategy of cooperative advantage as a means to compete. Morgan's trusts ate up the competition while Westinghouse partnered with it. Again, it shows the bias of the press, which tried to portray Westinghouse as a poor businessman. Westinghouse, however, had shown even more innovation in business and management than even in his engineering. Most of this can only be seen in hindsight that was not perceived by his earlier biographers.

80 Passer, 342

Chapter 14. Duopoly

Westinghouse's victories at the World Fair and later Niagara did not end the war of the currents. In many ways it only intensified it and spread it to other product lines, like motors, streetcars, and electrified trains. In 1895 little had changed, at 1 AM, you could find Morgan in his library plotting strategy for a victory over Westinghouse, and Westinghouse at Solitude or Erskine drawing new competitive designs. The two had come to a standoff. Westinghouse would never voluntarily come into the electrical trust, and Morgan needed to have more market control for General Electric to perform better. In 1895 there were only General Electric and Westinghouse operating in the electrical industry. It was a duopoly with lots of market competition, variation, and inefficiencies. It created many sleepless nights for Morgan and Coffin.

On the organizational front, Westinghouse had proved the better. His highly motivated engineers had bested General Electric time after time. Westinghouse's Pittsburgh lawyers had stood up to the best money could buy. His managers were masters of motivation, goal achievement, and employee satisfaction. His partners and suppliers demonstrated a type of loyalty that confused the best of the robber barons. Westinghouse's research and development model was world-class in every aspect. Morgan's advantage in banking had yet to penetrate Westinghouse's armor. Morgan had trophies on the wall, such as Thomson, Sprague, Edison, Forbes, Brush, Stanley, and many others only Westinghouse and Tesla roamed free. Morgan's control of the New York and even the Pittsburgh press had failed to down Westinghouse. Morgan's great merger artists, such as Villard, Adams, and Coffin had failed to pin Westinghouse. Westinghouse's ethics had trumped backroom deals. Even the economy had not defeated Westinghouse, but still he had only achieved a costly standoff, and Morgan was an extremely patient man.

The legal costs for both companies had been in the millions and it could be said that it cost the consumer as well. In a press release in 1896, General Electric estimated that General Electric had two-thirds of all the electrical patents and Westinghouse a third.[81] Westinghouse owned the patents of Sawyer-Man, Maxim, Tesla, Stanley, and Weston, while General Electric owned those of Thomson, Brush, Edison, Sprague, Bradley, and Van De Poele. Westinghouse controlled AC power generation and AC motors, while General Electric controlled most electrical equipment, incandescent lighting, and electric traction. In 1896 an amazing 300 lawsuits were in litigation covering all fields of electrical applications. Of course Westinghouse had General Electric blocked on the Tesla motors, and General Electric had blocked Westinghouse on some electric traction and incandescent lighting patents. Both Morgan and Westinghouse believed this litigation to be wasteful, although for different reasons. Morgan saw it as a classical failure of unrestricted capitalistic competition. Westinghouse saw it as a gross disrespect for intellectual rights. Westinghouse seems to have become more open to an agreement when he lost the trolley suit, which blocked his pet project of electric traction. These constant battles were also taking valuable time away from more productive work. Morgan was also convinced he had nothing with which to force Westinghouse into a merger after ten years of trying. Morgan could at least gain cost control, and maybe de facto control of industry prices. Westinghouse still had to contend with earlier attacks on Tesla's AC system, but these at least didn't have Morgan money behind them.

With the World's Fair and Niagara behind them, Westinghouse Electric and General Electric started to look at the cost of legal battling. Westinghouse was now focused on the electric traction business, and he was again finding stiff legal battling. In early 1896, Westinghouse lost a number of patent fights that would have ended his latest pet project of electric traction. Westinghouse now found himself in a position where a trade with General Electric made sense. The arrangement was announced on March 13, 1896. The two companies agreed to share patents and royalties under the management of a board, which became known as the Board of Patent Control. The Board consisted of four members, two from each company, and a fifth member to act as an arbitrator. Westinghouse Electric named George Westinghouse and Paul Cravath (his patent lawyer) with the alternatives being Charles Terry and B. Warren. General Electric named C. Coffin and F. Fish with Eugene Griffin and Gordon Abbott as alternatives. The fifth member was E. Thomas, who was never needed in the agreements fifteen-year existence. The government (McKinley administration) showed little interest in

81 "The End of the Electric Patent War," *Scientific American*, March 21, 1896

exploring its legality. Since General Electric and Westinghouse controlled the market, this was monopolistic behavior. Pressure did mount to the point that in 1911 both companies allowed the agreement to expire.

In 1896 Tesla still had not come to Niagara being obsessed with his wireless experiments, but in July of 1896, Westinghouse invited him and he accepted. Westinghouse had promised Tesla it would be a vacation, but he also planned a little business. Westinghouse sent his Pullman car to bring Tesla to Pittsburgh first. Tesla was entertained at Solitude by George and Marguerite, who had always been close to Tesla. Westinghouse was able to get caught up on the wireless, X-ray, and radio experiments. Westinghouse also spent a day with Tesla at the East Pittsburgh Works. He and Tesla had lunch at the Works with some old friends, such as Benjamin Lamme and Charles Scott (Scott had been Tesla's assistant in 1888 at the Garrison Alley plant). It became obvious that Tesla had moved beyond the world of motors and power generation, but he was happy to see old friends. Long after Westinghouse's death, Tesla would drop in at the East Pittsburgh lunchroom to visit friends, but this 1896 visit was his first to the plant. On July 19, 1897, Westinghouse and Tesla put together a group to visit Niagara. The group included Westinghouse's young son (thirteen at the time); Paul Cravath, Westinghouse's New York lawyer; Thomas Sly, manager of electromotive power for the Pennsylvania Railroad; George Melville, US Navy; and Edward Adams and William Rankine of Cataract Construction. Thomas Sly had been invited as part of Westinghouse interest in electric trains, and Commodore Melville represented Westinghouse new interest in marine turbines. Tesla seemed awed at the gargantuan turbines and generators. Tesla was always more scientist than engineer, and preferred the laboratory to giant pieces of equipment. Actually, Tesla had a bit of a phobia of large electrical equipment believing it affected his back and legs. Tesla gazed at the huge white Hydro-Electric Plant designed by the famous architect Stanford White, noting it lived up to its name-'The Cathedral of Power." Tesla's favorite part of the trip was lunch at the Cataract Hotel, which overlooked the Falls.

In July of 1896, the power of Niagara was supplying a massive electrochemical industry at the Falls, but had yet to be yoked for use at Buffalo some twenty-six miles away. The electrochemical plants had been receiving power for almost a year, but Niagara power was originally built for the transmission of power to Buffalo. It was the event that the press and Westinghouse had been waiting for. On November 15, 1896, the Niagara Falls Power Company would take 2200 volt current from the Westinghouse/Tesla dynamo and using a General Electric transformer stepped the voltage up to 10,700 volts for transmission to Buffalo, twenty-six miles away. It would also be changed from two-phase to polyphase by the GE transformer. The Cataract Power and Conduit Company would receive the current and step it down for use in street arc lights, incandescent lights in homes, electric motors in factories, and to the Buffalo Railway Company for streetcars. Of course, for streetcars rotary converters would change it to DC current. This was the victory that Westinghouse had really been waiting for.

The final celebration of the yoking of Niagara was planned for January 12, 1897. This would be the academy award of 19th century industrialists, scientists, and engineers. Over 350 were invited to this ornate event at the Ellicott Club of Buffalo. The Ellicott Club was on the top of the ten-story Ellicott Square Building, said to have been the world's largest office building of the time. Thomas Edison chose not to come, probably because the AC power plant represented one of his failed predictions. Francis Stetson, Morgan's lawyer, acted as the host although both Morgan and John Astor did not come. Nikola Tesla was asked to be the speaker of the event. Many had wanted Westinghouse to be the speaker, but he feared public speaking and avoided it whenever possible. Morgan's absence might well have been related to the selection of Tesla as the speaker. Each quest would receive an engraved aluminum covered program, the aluminum having been produced by the Pittsburgh Reduction Company at the Falls. Tesla's speech would characterize Niagara as a utopian dream and a triumph of capitalism:

> Among all these many departments of research, there is one which is of the greatest significance for the comfort and existence, of mankind, and that is the electrical transmission of power . . . We have many a monument of the past ages exemplifying the greatness of nations, power of men, love of art and religious devotion. But that monument at Niagara has something of its own, worthy of the scientific age, a true monument of enlightenment and of peace. It signifies the subjugation of natural forces to the service of man, the discontinuance of barbarous methods, the relieving of millions from want and suffering . . . Power is our mainstay, the primary source of our many-sided energies.[82]

Turning the power on to the city of Buffalo was what Westinghouse had been working towards, not because he wanted to sell more power generation, but because he had positioned himself for the electrical equipment market. In early 1890, Westinghouse Electric had AC motors on the market. Westinghouse had 25 to 250 horsepower electric motors on the market. In 1891 Westinghouse electric motors had replaced some steam engines in running a steel rolling mill. Westinghouse, while at a disadvantage in DC streetcar motors had the lead in AC motors, thanks to the early work of Tesla at Westinghouse Electric in1888. The Westinghouse general-purpose electric motor had advantages of better torque for tough applications like cranes, and a significant cost advantage because of the manufacturing efficiencies at East Pittsburgh. While the patent agreement had eliminated the legal advantage of Westinghouse over General Electric, it did not prevent Stanley Electric from producing AC motors in direct violation of the Westinghouse owned Tesla patents. The Stanley case remained tied up in an extended legal battle. Still the exponential growth of the electrical motor and electrical equipment reaped profits enough for all. From 1890 to 1900 alone horsepower usage for mechanical power went from 16,000 horsepower by electricity

82 Nikola Tesla, "Niagara Falls Speech," *Electrical World*, February 6, 1897

out of a total of 5,955,000 horsepower to 495,000 out of a total of 11,300,000 horsepower.[83]

The commercialization of Niagara and the Patent Board Agreement finally freed Westinghouse to explore new areas. One of these was his boyhood toy of a rotary steam engine, and another was an engine powered by natural or coal gas. The gas engine seemed to Westinghouse to offer a replacement for the steam engine of the Gilded Age, but even as Westinghouse pioneered the electric motor, which would be the demise of the steam engine, he worked on a gas engine. The gas engine ran on a mixture of fuel gas (or natural gas) as a horizontal internal combustion engine. French inventor Etienne Lenoir made the gas engine a commercial success in Europe in the 1870s. The horizontal internal gas combustion engine was based on a four-cycle sequence, similar to today gasoline engine (gasoline is vaporized to a gas in our car engine). Nikolaus August Otto improved on this engine and Otto found applications in factories. By the early 1880s, gas engines could produce 20 horsepower. Westinghouse and Westinghouse Machine Company started to sell gas engines in 1880s to augment the natural gas business of Westinghouse's Philadelphia Company. These small engines could not do the work of steam engines, but Westinghouse pursued the business because of the high thermal efficiency of the gas engine. The gas engine applied fuel directly to mechanical work, while steam engine used coal to create steam to drive mechanical pistons. It was the type of engineering project Westinghouse loved that of efficiency. From Westinghouse Machine Company's earliest days, it had focused on efficiency. Westinghouse Machine Company was always happy to send an engineer to a steam plant to facilitate improved efficiency points. Westinghouse Machine printed pamphlets on steam efficiency as well as publishing endless journal articles on efficiency.

The efficiency of the gas engine was too good to ignore. In the early 1890s, the gas engine had several problems. First was its variable speed resulting from no good method of regulation and second was its low horsepower. The variable speed rendered the gas engine useless for electric lighting. Westinghouse had used the gas engine as a project for relaxing during many train trips. Westinghouse found relaxation during the World's Fair preparation by making engineering designs and drawing of a gas engine regulator, often this was done on his seemingly endless railroad commuting between Chicago, New York, Pittsburgh, Washington, and Berkshire Hills. The Westinghouse regulator used a flywheel arrangement to smooth out energy delivery between intermittent explosions. This patent successfully resulted in a controlled speed engine that could be used for electric power generation for electric lighting. In 1896 with the war of the currents over, he returned to building even larger gas engines. For a time he even envisioned large gas engine power stations for electrical generation. This dream started another major research project of Westinghouse Machine Company in the develop-

83 Passer, 343

ment of synthetic fuel gas to assure steady prices and supply. In 1896 *Scientific American* shared his enthusiasm:

> Of scarcely less importance is the question of the increased consumption of the gas resulting from the extended use in gas engines. For some reason or other the gas engine has not received the attention in the United States that it has in England. No doubt the price of gas here has been prohibitory; and greater attention that has been paid in this country to the development of the electric motor has caused the gas engine to be neglected. It is stated that there are many European manufacturers who will furnish gas engines guaranteed to run onl7 cubic feet of gas per horsepower per hour; and Mr. Westinghouse, in this country, is building engines, which he will guarantee to use 20 cubic feet per horsepower hour. If gas could be supplied at the Manchester price (60 cents per thousand), such an engine would cost twelve cents per horsepower per day for a day of ten hours. If the figures of Mr. Westinghouse's guarantee can be realized in practice, and if there are no local conditions, which make it impossible to manufacture gas in the United States at something less than $1 per thousand feet, there should be a great future for the gas engine in this country.

Westinghouse believed at one point that gas engine central power stations might spread across the country; however, he was able to reach only 500 horsepower for most applications. Westinghouse went even further, calculating that gas engines generating electricity would propel electric trains. In an 1896 Scientific American article Westinghouse prophesized:

> The Pennsylvania Railroad today, it is said, consumes about 5,000,000 tons of coal per annum on its lines east of Pittsburg, taking, approximately, 20 loaded trains each day for its transportation, and consequently the return of 20 empty trains, and requiring for the service of the company alone fully 3,000 cars and a proportionate number of locomotives. If this were to be generated by gas engines, only about one-eighth, or 600,000 tons of coal per year, would be required, effecting a saving of over 4,000,000 tons of coal, now costing the railway company above $5,000,000 – a saving which would justify a large enough capital expenditure to cover the complete equipment of the railway. To carry out an arrangement of this character, stations having electric generating plants with gas engines and producers could be located at intervals of from ten to twelve miles, so that there would always be two or three stations furnishing current for particular part of the line.[84]

Westinghouse did move into the production of gas engines with a good deal of success. In particular, 300 horsepower applications became popular for localized lighting. By 1901 Westinghouse gas engines approached 500 horsepower and were used to light part of the Pan-American Exposition at Buffalo. Also to a large degree, he was competing internally with the rapid developments of steam engines by Westinghouse Machine Company. Westinghouse soon moved from gas engine development back to the ideal of the steam turbine, which could achieve much higher horsepower for electrical generation.

84 'Gas Engine Stations for Trunk Line Railways," *Scientific American*, February 15, 1896

In 1896 Westinghouse Machine was expanding into large Corliss steam engines for power generation. The steam turbine was also becoming one of the challenges to all types engines, and this also fit with his earliest work on the rotary engine. As more stories of steam turbines appeared Westinghouse turned to one of his earliest passions. As noted, Westinghouse's first rotary engine patent of 1865 had reaped no commercial profits. The boyhood models could now be found in the basement at Solitude and well as some of Westinghouse Electric's labs. Irish engineer Charles Algernon Parsons was pursuing the dream of a rotary steam engine passionately as Westinghouse was locked in the war of the currents. Parsons developed a steam turbine for ship lighting. In 1884 parsons used his experimental ten horsepower turbine to power lighting on a yacht called *Turbinia.* Parsons even tried to use his turbine to propel ships. The reports of Parsons turbine reminded Westinghouse of his propelled ship model he had developed while in the Navy. In 1885 Parsons patented the steam turbine, but it was little better than Westinghouse's rotary engine. Parsons, however, continued his developmental work throughout the 1880s. He founded a company for the production of steam turbines, and by 1889 he some 300 turbines of 75-kilowatt capacity. By 1894 Parsons had a 300-kilowatt turbine generator. Upon hearing that Edison had done some earlier experimental work with turbines, Westinghouse moved to purchase the patents and licenses needed to manufacture steam turbines in the United States in 1896. Actually, General Electric had already been to England to study the Parsons turbine for purchase. Charles Steinmetz, General Electric's brilliant electrical engineer, felt the Parsons's devices too crude for development. Steinmetz reported the following to E. Rice, Vice-President of GE Engineering in 1895 Referring to the question of steam turbines, I consider this field very promising, and believe we should push experimental work in this regard with the greatest possible speed to see what we can do in this matter. I do not think very much of Parsons's steam turbine but think that a simpler design even if not quite as efficient in steam consumption, would be desirable . . . I prefer to delay any agreement with Mr. Parsons."[85]

It would take the visionary foresight of Westinghouse to see the potential of Parsons rough designs. Westinghouse sent E. Keller, Vice-President of Westinghouse Machine, to purchase or make a licensing agreement for these Parsons patents, which like the Tesla electrical would help Westinghouse start another engineering revolution. Keller also initially hesitated, feeling the Parsons were not fully commercial, and had too many problems at that time. In retrospect, E. Keeler years later would state, "None of us fully realized then that this agreement was the death knell not only of Westinghouse's own rotary hobby, but also of the large reciprocating steam engine.

85 Letter from C. Steinmetz to E. Rice, January 3, 1895

It was an outstanding example of Westinghouse's farsighted vision of future needs and his indomitable courage in pioneering in new fields."[86]

Westinghouse soon found the Parsons turbine to be limited in application. Westinghouse coordinated a research project between Westinghouse Machine Company and Westinghouse Electric to improve the efficiency of the Parsons turbine. In 1898 Westinghouse brought in Edwin Musser Herr to head up the implementation of the steam turbo-generator. Herr would actually be hired as assistant manager for Westinghouse Air Brake, which was to become the first user of the turbo-generator. While Herr was a graduate mechanical engineer from Yale, he had both a strong railroad and electric background. Ultimately, Herr would become President of Westinghouse Electric in 1911. Steam turbines appeared now to him as the real future of power generation, and he shared the enthusiasm of Westinghouse. There could be no doubt about the passion of Westinghouse for this project. Years later Herr reported the following Whether he was in Pittsburgh or New York or Lenox, he invariably called me on the telephone several times a day to inquire how things were going."[87] Biographer Henry Prout noted a similar approach as the electric rotor insulation of the turbo-generator was being worked on Westinghouse was personally much interested in this part of the construction, and would telephone almost every day asking whether anything satisfactory or promising had been worked out. He was very prolific in suggestions for rotor construction, but, not being experienced in the difficulties of insulation, the engineers had to turn down his suggestions daily. However, as good reasons were given he took it all good-naturedly." Westinghouse took steam turbine generators to a state of commercialization once thought impossible. In 1898 Westinghouse installed three 300-kilowatt turbo-generator for his Wilmerding Air Brake plant. These replaced Westinghouse reciprocating steam engines being used for electrical power generation. The zeal for the development of the turbo-generator would carry through to the end of his career at Westinghouse Electric and Manufacturing Company.

The 1890s saw the evolution of Westinghouse as an American genius on a level with Edison. In mid-January of 1897, the Westinghouse's opened their third mansion on Dupont Circle in Washington D. C., known as Blaine Mansion. Blaine mansion had been built by James Blaine, three time Republican presidential candidate, in the fashionable Dupont Circle. While they planned for it to be their winter home, Westinghouse close working relationship with Senator and President McKinley often brought him to Washington. Marguerite, of course, had been advised by doctors to minimize her time in the sulfurous air of Pittsburgh. Blaine House was said to rival the best mansions in the city. Both George and Marguerite seemed to find the social life of Washington much to their liking. Their house warming party was the talk of the city with rose decorations throughout the mansion. It was the begin-

86 George Westinghouse Commemoration (New York: American Society of Mechanical Engineers, 1937)
87 Prout, 100

ning of the Washington social season introducing the city's new debutantes. Reporter Janet Jennings noted the celebrity status of the debutantes, "among the season's debutantes asked by Mrs. Westinghouse to assist were Miss Sartoris, General Grant's granddaughter, and the misses Sheridan, daughters of General Sheridan. They carried lovely flowers, presented to the hostess, who, like the late Mrs. Whitney, is a woman to get much pleasure out of life by giving pleasure to others."[88]

1897 was a time of relaxation for Westinghouse, and he spent more time than in previous years at his summer home Erskine in Berkshire Hills. Westinghouse put on his usual fireworks display, but added a light show that included a profile of President McKinley. Westinghouse had worked with Senator McKinley of Ohio on a number of railroad bills, and they had become very close. In late September his old friend President William McKinley visited him and Mrs. Westinghouse at Erskine. McKinley would visit Westinghouse again as he loved to vacation in the Berkshire Hills area. Prior to this in late August, Westinghouse made a trip back to Pittsburgh and Solitude to meet with Lord Kelvin. Westinghouse, Lord Kelvin, and Tesla also met in New York. Kevin had become extremely interested in Tesla's wireless experiments and Mars project. Lord Kelvin would stay a few days and tour Westinghouse's East Pittsburgh Works in September. Kelvin went on touring, but came to Erskine in October to spend more time with Westinghouse. Westinghouse showed little interest in wireless communications, but did discuss his dream of an electromagnetic perpetual engine.

Lord Kelvin and Westinghouse spent hours into the early morning going over many of Westinghouse's more marginal projects. They had for years been corresponding, but now they could exchange ideas rapidly. One of these Westinghouse curiosities dealt with the famous Second Law of Thermodynamics, which he had often conversed with Kelvin about. Kelvin had predicted in 1847, the possibility of extracting heat from air to heat buildings (known as a heat pump today). Westinghouse had read everything he could on Kelvin's studies. Westinghouse had another dream of harnessing power from the air to run equipment. Lord Kelvin spent hours trying to demonstrate the errors in this approach, but Westinghouse was an engineer, not a scientist, and went ahead with his experiments. Westinghouse in doing so would set the groundwork for air conditioning, refrigeration, and heat pump equipment. Biographers and writers have made much of Westinghouse's hardheadness in refusing to stop experimentation on the second law "exceptions." Engineers of the time used it to point to the fact that Westinghouse was not a "trained engineer." Others used to refer to his inability to listen or apply science correctly, but all this really misses the point. In Westinghouse's experimentation, we see the purest form of engineering. An engineer must know science, but he can never be limited by it. Westinghouse respected science and great scientists, but he would not allow himself to be limited

88 Janet Jennings, "Our Washington Letter," *The Independent*, January 21, 1897

by either. Had he done so he would have bowed to Vanderbilt's opinion of the air brake, Edison's opinion of AC motors, and the International Committee of scientists on Niagara. In a way, Westinghouse was always that little boy that wanted his way, but now he had learned to direct rejection into goal setting and hard work. Henry Prout put it perfectly:

> What has been told is an extreme example of Westinghouse's independence of mind. He was not "entirely satisfied without making the experiments." He accepted nothing on a great name or a great position. But it must not be inferred that he was lightly skeptical. Far from it, he was a reverent man in mind and soul. This was his attitude toward religion, toward the state, toward the courts, toward the family, and toward his father and mother. He respected established things; he revered high and fine things. This was not a matter of reason, but of instinct.

Prout's analysis helps shine some light on some of the inconsistencies and paradoxical actions of Westinghouse. As a child he wanted his own way often, and as a learner he demanded experimental proof. He was in his heart a rebel, but one with deep Christian values passed on from his parents. He learned to deal with being told no in positive ways, some might even say obsessive ways. Some of his technical endeavors were personal power drives to prove himself right. He was a workaholic driven to achieve. The root of his persistence was often the desire to prove the experts wrong. Achievement was often his revenge, but at the same time failure looms as his Achilles heel.

Chapter 15. Industrial Diplomat

> "No, I do not feel that it would be right to stop here; I feel that I have been given certain powers to create and develop enterprises that other men can find useful and profitable employment"
>
> — George Westinghouse

By the late 1890s, Westinghouse was a national celebrity. Still, few in his hometown of Pittsburgh knew much about him. While he avoided being photographed, his 6 foot 3 inch frame and famous walrus mustache made it easy for those in the know to recognize him. The long dark overcoat and top hat were also very characteristic of the man. He was always in the shadow of Andrew Carnegie, Charles Schwab, and "the boys of Braddock" around Pittsburgh. Westinghouse refused to have his name used on buildings and alike. He also avoided having his name in print for any charity. For example, in 1899 Andrew Carnegie made headlines in the Pittsburgh papers with his $20,000 donation for the famous Allegheny Observatory. Westinghouse quietly donated a complete electric plant and installation.[89] Even today Pittsburgh tour guides can point out such Carnegie donations, but are at a loss to even know of Westinghouse's. The same was true of Westinghouse's technical achievements. At Niagara he gave credit to Tesla, trying always to get to the back of the room. Some of his greatest humanitarian accomplishments such as the town of Wilmerding, remained almost unknown to neighboring Pittsburgh citizens.

---

89 "The Allegheny Observatory Objective," *Scientific American*, September 2, 1899

Wilmerding had however, captured the interest of many politicians around the world. A young Prince Albert of Belgium made a tour of the United States in 1898. He was first entertained at the McKinley White House where he inquired of Mr. Westinghouse's revolutionary ideas in industrial relations. McKinley being a friend of Westinghouse made arrangements for the Prince to visit Westinghouse in Pittsburgh. The Prince traveled directly to the Pennsylvania Railroad station at Westinghouse's Solitude using personal railroad car —Glen Eyre. They had an extensive dinner at Solitude before leaving the next day to view Wilmerding. On the way from Solitude to Wilmerding, they passed by the First Ward slums of Carnegie's Braddock. The site of such industrial slums offered the contrast that the Prince had heard about. The most notable remembrance of the Prince was a bronze plaque with a Westinghouse quote I want you to know and feel that no one has your best interests at heart more than I have. We are all interested in the same object, to make this company a success. I have my part in the job and you have yours and, if we all work together, in friendly cooperation and with a feeding of mutual good will and good fellowship, the desired result can never be in doubt." Prince Albert would go back to Europe an advocate of improved industrial relations and social programs.

Westinghouse had many opportunities to return favors to McKinley. One concern was that the feud between Andrew Carnegie and Henry Clay Frick might cause McKinley and the Republicans to lose the election in 1900. The very public feud between the two had come back to the Homestead Strike of 1893. Frick was then in charge of Carnegie's steel mills while he was in Scotland. Carnegie publicly blamed Frick for the labor problems at Homestead, and a battle started that would eventually result in Frick's ouster as a Carnegie Steel director. In 1898 Frick filed suit claiming he was forced out and not compensated for his shares. The affair was bringing a lot of the dirt of Big Business and the Republicans to the front page. Westinghouse, having been a friend of both, was asked to try to broker a peace. From Blaine House Westinghouse sent a letter calling it, "a calamity by reason of the fact that the private affairs of your company will undoubtedly be made public . . . I may add that Mr. Frick has recently spoken to me in such terms that I feel there must be a way to adjust matters between you and him." Westinghouse, while a friend of Frick, was more moved by his support of President McKinley. Other industrialists joined Westinghouse and ultimately the suit and the public battle ended.

Westinghouse missed few opportunities in his life, but like most giants in their fields, age brought conservatism in his views. Wireless communications would be one of those rare events, and had Westinghouse jumped into wireless, our grandparents might have been using cell phones. Nikola Tesla might be the exception in that his dreams and projects started to border on science fiction. In 1899 (following in part his dreams to communicate with the planets such as Mars) at the request of Leonard Curtis, a Westinghouse lawyer, Tesla built an experimental station in Colorado Springs to experiment in wireless communications. Curtis had agreed to supply free electrical power through El Paso Electric Company. Tesla's station was two miles up in the Rocky Mountains. Tesla started to truly emulate Doctor Frankenstein with a number of far-fetched experiments, many of which would not be known until a notebook was discovered in the 1950s. He ventured into all types of curious phenomena and the occult, but much of it resulted some practical advances in wireless and radio transmission.

Tesla's success with wireless transmission moved him to propose a wireless transatlantic program to compete with the transatlantic cable. At the end of 1899, Tesla had been writing Westinghouse about the possibility of this project. It was hard for Westinghouse to say no to Tesla, who had been integral in the success of Westinghouse Electric. In January, 1900 Tesla came to Pittsburgh for an evening at Solitude and a ride on Westinghouse's Pullman car to New York the next day. Westinghouse had established an office in New York several years before and would make the trip several times a month. Tesla asked for $6000 dollars and hoped further that Westinghouse would invest more in the venture. Tesla was strapped for cash because of his high living, but Westinghouse was also strapped. Most of Westinghouse's wealth was in his companies, and Westinghouse Electric continued to require high cash layouts. One author believed Westinghouse Electric to be overextended by $70 million[90], but the annual report suggests half of that. Some of this was the result of Westinghouse's continued various lawsuits over Tesla's polyphase patents. Westinghouse would lend Tesla the $6000 as a personal loan, but suggested Tesla try New York bankers for capital. Tesla would ultimately get J. P. Morgan involved and probably Astor. The story has an amazing similarity to the war of the currents with patents, with Marconi, but this time Westinghouse was on the sidelines.

The battle for patent priorities had continued long after the 1896 patent agreement of Westinghouse and General Electric. Several old Westinghouse opponents, such as William Stanley, Elihu Thomson, Walter Bailey, and Charles Bradley continued their lawsuits against both General Electric and Westinghouse Electric. These lawsuits went back to Tesla's earliest work and that of some European scientists. Westinghouse, as the owner of the Tesla AC patents, was forced to continue the legal battles. Most people, including Tesla himself, credited the French astronomer Francois Arago with the discovery of rotating magnetic fields in 1824. The battle revolved around whether British inventor Walter Bailey or French inventor Marcel Deprez had discovered the AC motor's application to rotating fields. It would not be until Westinghouse had passed on the Tesla venture that, at the end of January, a court ruling would at least free Westinghouse of these legal costs. The Southern circuit US court rendered what the *Pittsburgh Dispatch* called the "doctrine of obviousness."[91] Judge Townsend ended almost twenty years of legal fighting stating:

---

90 Seifer, 238

91 *Pittsburgh Dispatch*, February 23, 1901

> The apparent simplicity of a device often leads an inexperienced person to think it would have occurred to anyone familiar with the subject, but the decisive answer is that though dozens and perhaps hundreds of others had been laboring in the same field, it had never occurred to anyone before . . . Baily and others did not discover the Tesla invention; they were discussing electric light machines with commutators . . . Eminent electricians united in the view that by reason of reversals of direction and the rapidity of alternations, an alternating current motor was impracticable, and the future belonged to the commutated continuous current . . . It remained for the genius of Tesla to . . . transform the toy of Arago into an engine of power.[92]

This finally ended another cash drain on Westinghouse Electric, but it was probably too late to help the slide into the Panic of 1907. Tesla, other the other hand, was into a new line of wireless experiments needing money badly. He was building a laboratory in Colorado to perform experiments including trying to communicate with life on Mars. Tesla did not pressure Westinghouse for money, but did proceed to order special generators from Westinghouse Electric and contracted Westinghouse, Church, Kerr, and Company to build two power plants. Westinghouse helped by assigning an engineer full time to Tesla's project. Tesla worked with Morgan, Astor, and Frick for money that came in slowly. Ultimately, the project died a slow death and Westinghouse's organizations never received full payment till after 1912.

Going into the new century, Westinghouse was focused on the development of the turbo-generator, marine turbine, and electrifying railroads. These were major technical challenges, but they were not his only. The turbo-generator was probably his most successful, and would promote Westinghouse Electric to industrial dominance for decades. As seen, the work goes back to his earliest work on the steam rotary engine. With the purchase of the Parsons's patents, Westinghouse's interest was renewed in 1896. Westinghouse, as with his experience with the Gaulard and Gibbs transformer, took this crude turbine and turned it into a commercial system. The experimental units at Air Brake only served to inspire Westinghouse more, devoting more of his limited energy to the project. Westinghouse was not satisfied with the operation. The experimental units at the Air Brake power station were 300-kilowatt generators and were operating in 1898. The competing units using reciprocating engine generators were being made that could generate 6000 kilowatts, and the average of 3500 kilowatts was common. Westinghouse reciprocating engines were driving these generators as well as those of General Electric, so the market pressure was not immediate. In fact, many at Westinghouse Machine were questioning the business logic of the thrust. Passion was driving this project for Westinghouse. Westinghouse faced a maze of technical problems to close the gap between the large reciprocating engines of the period.

The same year, Westinghouse launched a project to build a 1500-kilowatt (1200 revolutions per minute) turbo-generator. The turbo-generator

92 "Tesla Patent Decision," *Electrical World*, May, 1902

would become known as the "rotating field type" because the windings were tucked in the turbine rotor. The project was one of George's "pet projects" and commanded daily involvement. Years later in 1939, Frank Smith, the President of Consolidated Edison (he had worked with Westinghouse in the turbo-generator development), called the project "one of the great romances in the career of George Westinghouse." It was the project that Westinghouse would put beyond the World Fair and the war of the currents. Many train trips from Solitude to east Pittsburgh ended in frustration as bearings failed and noise levels became unbearable. Westinghouse moved Benjamin Lamme to take over the new design project to eliminate the noise. Lamme noted the radical reversal of Westinghouse to develop the turbo-generator It seemed a pity, after I had spent so many years in the development of engine type generators [reciprocating], that I should have to turn in and help undo this work. However, it has always been my practice to follow up what I considered the most promising line of endeavor; and as the turbo-generator held out greater promise than the engine type."

Lamme designed in the necessary improvements, but forged rotors and shafts became a problem. They first tried forgings from the great iron works of the Krupp in Germany, but the shipped in forgings failed to meet quality control tests. Westinghouse engineers moved to cut steel plates, against the popular belief that steel was not a good rotor material. Finally Westinghouse and Lamme came to agreement to try steel castings. This use of steel castings was a radical design step for the time. Westinghouse was, however, confident in steel casting, and as a young man had pioneered its use in railroad applications. The failure resulted in a study and ultimately improvements were mandated by Westinghouse. The danger of castings was a defect that caused explosion under rotation stress. Westinghouse and his young engineer, Frank Smith, had only a few years before experienced an explosion of a cylinder head in a large reciprocating engine, which caused parts to fly through broiler plate as Westinghouse and the others ran for the streets. These radical shifts in direction in organizational focus based on the best technology were one of the winning attributes of Westinghouse's style.

During the frenzy of the development work, General Electric won the contract for the Manhattan Elevated Railway. This used seventeen reciprocating engines to develop 6,000 kilowatts each (rated at 7,500 kilowatts). It was the largest electrical contract ever, and while General Electric won the contract, Westinghouse reciprocating engines were the drivers. Few realized at the time that the Manhattan Elevated would represent the peak of the reciprocating steam engine power generation. These Manhattan engines would indeed be the largest ever built of that engine type weighing nearly a million pounds each. Westinghouse persisted in his development of the turbo-generator and in 1900 installed the first commercial application of the 1500-kilowatt generator for Hartford Electric Light Company. This Hartford turbine would be four times more powerful than any turbine in the world at the time.[93] Westinghouse had to make major adjustments to the East Pittsburgh plant to manufacture these huge turbo-generators. The actual kilowatt increases for turbo-generators came slowly, but the advantages were quickly realized. The turbo-generator could operate as high as twice the efficiency of a reciprocating steam engine. Remember that the reciprocating engine had to transform vertical motion into horizontal. The maintenance costs which long run could be more significant than the capital cost of the engine, were only a fifth of the cost of the reciprocating steam engine.[94] This reduction in maintenance cost was due to the fact that a turbine could run months without attention, while a steam engine must be adjusted, oiled, and inspected very often. The engineering constraints on the reciprocating engines limited to a maximum of 10,000 kilowatt, while by 1907, 10,000-kilowatt turbo-generators were being built. That reached 30,000 kilowatts by 1917, and 200,000 by 1935. Other advantages included reduced power station space with the turbine requiring less than 8% of the space required for reciprocating engines. In 1935 the savings from turbo-generators over reciprocating engines was $35 million a year. Interestingly, General Electric again found itself years behind Westinghouse and with an inferior design.

If Westinghouse had built his own monument, his primary achievement would have been that of the turbo-generator. It was a project that spanned his life, starting at his father's workshop with a simple toy model. Henry Prout chronicled the radical change in power generation by the turbo-generator:

> In consequence, this business [reciprocating engines] went to almost nothing within practically a year's time, and, in fact, the engine type [reciprocating] went down before the turbo-generator was ready to take its place. This obsolescence of the engine type alternator [generator] was almost pitiful. Here was a branch of heavy engineering, built up at great cost and backed by years of experience. In coming of the turbo-generator this experience was almost thrown away, for the engineering required in turbo-generator work was so radically different from that of the engine type generator that designers had to start practically anew and build up entirely new experiences at enormous expense and years of effort.

It was the expense that would cost Westinghouse Electric so much in this pioneering effort. While engineers and society in general hailed the accomplishment, bankers looked on in fear of the development costs. Frank Smith at the commemoration of Westinghouse's 90th birthday (1938) said it best the world had to wait 2000 years for the genius of George Westinghouse to visualize that the offspring of a union between the plaything of Hero and the alternating-current generator would become the universal burden bearer of mankind."

In the midst of his development of the turbo-generator, Westinghouse traveled to Europe for several months in 1901. This trip was to address the financial losses in Europe that had been draining Westinghouse Electric in East Pittsburgh. Europe in general had been a mixed bag of results for West-

93 "George Westinghouse," George Westinghouse Museum Foundation
94 Passer, 312

inghouse, but electrical power equipment provided one of the first international markets requiring a world presence. Westinghouse had first expanded his brake business to England in 1870s, and by 1889, he had a Westinghouse Electric Company known as the London Company. British Westinghouse Electric and Machine was formed in 1899 with the opening of a new plant in Manchester. The great plant was built with all the efficiency of the East Pittsburgh, but the company would never be profitable. The problem centered around the political power of England's coal industry and railroad industry, which fought the conversion to electricity. British newspapers called it "the invasion of England." By 1901 the home base of Westinghouse Electric could no longer manage the financial drain, and Westinghouse needed to do something. In 1901 Westinghouse had come to England to settle some local financing as well as bid on a turbo-generator contract. Westinghouse was able to win a contract to build 8,000-kilowatt turbo-generators for the Clyde Valley Electrical Power Company. This type of engineering approach became a weakness for Westinghouse. Westinghouse, in hindsight, probably should have gotten out of Britain. The losses were too great, and the political battles endless. It's hard to tell whether any businessman would have proceeded differently. The electrification of the Clyde Valley offered huge potential; the Clyde Valley of Scotland was one of Britain's most highly industrialized area.

Westinghouse's plan for Britain was to build power plants near the coalfields and distribute AC electricity throughout the nation. The British railroads, however, were now codependent on distributing coal to local power stations. The reduction of coal shipping would damage their profitability, and they used all their considerable political power to oppose Westinghouse. The distribution of coal by railroads was an extremely inefficient overall means of energy distribution. Westinghouse's plan would be slowed at every turn, costing Westinghouse Electric millions, but here Westinghouse's steadfastness worked against him. He was right and correct, but the political environment was not ready. Years later, Westinghouse lawyer, Paul Cravath, reported a British leader as noting Mr. Westinghouse's conception of what should have been done was faultless. It was his misfortune that was a quarter of a century ahead of the times. If Great Britain had accepted his advice countless millions of waste would have been saved. It will now be necessary to scrap enormous investments in un-economical plants to make way for the carrying out of Mr. Westinghouse's plan." In fact, the British Electricity Bill of 1919 would confirm the soundness of the Westinghouse plan, but unfortunately, it was too late to save Westinghouse Electric from financial collapse.

Another reason for Westinghouse's 1901 European trip was the reorganization of all his brake and electric companies. He traveled to Paris, which had, had an Air Brake Company since 1879. His European operations in general lacked the strong organization of his American operations. The European efforts deserve some criticism. First, the organizations lacked strong local control, preferring to centralize strategic management under American control; at the same time, tactical control was also confused with the French reporting to the British company. Westinghouse had started the building of his British Manchester plant in 1901, and used as many local contractors as possible. He wanted to eliminate the "invading army" image that the press was highlighting. In an earlier project in 1888, Westinghouse had built a power plant for London with everything being manufactured in Pittsburgh and shipped over. The problem for the Manchester plant had been the long construction time estimates by British contractors. Westinghouse decided to hire James Stewart of New York to manage the project with the stipulation that only British labor would be used. This allowed him to shorten the timeframe using American methods. The project required 2500 laborers, 12 million feet of British lumber, 10 million British bricks, 15,000 tons of British steel, 175,000 feet of British glass and 40,000 feet of paving. The economic influx helped mute the screams of a "Westinghouse Invasion of England." The use of local suppliers and labor became a model for later companies to use in international expansion.

Westinghouse was also coming into direct competition with Rothschild backed European companies in the manufacture of electrical equipment. Westinghouse's American banker, August Belmont, really represented an American branch of Rothschild. In fairness, Westinghouse pioneered the rise of the American international companies lacking prior experience. Westinghouse's French operations had been one of his profitable organizations. Westinghouse tried to resolve the problems by bringing operations in France, Belgium, Spain, Portugal, Italy, and Holland under the newly formed Societe Anonyme Westinghouse. The new French based company was successful, but the huge capital drains of British Westinghouse continued. While in Paris Westinghouse was able to meet with the famous French engineer Maurice Leblanc. Leblanc had invented an air pump condenser system that saved steam and greatly improved the efficiency of the steam turbine. Leblanc also had some generator devices of interest to Westinghouse that had been studied by Benjamin Lamme in East Pittsburgh. Meeting at Westinghouse's hotel on the rue de l'Arcade, Leblanc remembered the fairness, mechanical expertise, and humor of Westinghouse. Months later, Westinghouse, not only bought the American rights, but also hired Leblanc for the French operation. Before leaving Europe, The Westinghouse Electricitats-Aktiengesellschaft was organized in Berlin.

On Westinghouse's return from Europe in 1901, he beefed up three new projects that of the marine turbine, the Nernst lamp, and electrifying railroads. The fall of 1901 marked the formation of the Nernst Lamp Company in Pittsburgh. Nernst Lamp was a major personal investment for Westinghouse, as Westinghouse Electric could not increase debt for such investments. The Nernst Lamp had huge potential. Westinghouse had once called the incandescent lamp the most inefficient appliance in the world. 90% of the energy of the coal used was lost with incandescent lamps. In engineering terms the incandescent lamp used four watts per one candlepower, while the Nernst Lamp used only a half-watt per candlepower. The Nernst Lamp was a forerunner of the tungsten lamp of today. The Lamp was the invention of Dr. Walter Nernst, a German physicist. Westinghouse had seen the invention on an earlier trip to Germany, and in 1898 had asked Dr. Nernst to come to East Pittsburgh. The filament of this lamp was a porcelain rod made of rare earths such as yttrium. Rare earth filaments could outlast the best tungsten ones available. This required Westinghouse to purchase a rare earths mine in Barringer Hill, Texas. The lamp immediately caught Westinghouse's attention when tests showed it to have twice the efficiency of most incandescent lights. The glow of the lamp was considered to be close to natural sunlight. Westinghouse purchased the American rights and formed a development team of Alexander Wurts, Henry Noel Potter, and three other electricians. Returning from Europe he formed the Nernst Lamp Company and opened a factory at Garrison Place and Fayette Street in Pittsburgh. The bankers were now becoming considered about the debt position of Westinghouse Electric, forcing Westinghouse to personally invest in Nernst Lamp.

The lamp had made its début at the Buffalo Pan-American Exposition in 1901 with a resultant boom in sales. Sales peaked at about 130,000 Nernst lamps in 1904, and then started falling to sales of improved tungsten incandescent lamps. Ultimately Nernst Lamp Company would go under in the Panic of 1907. Westinghouse experimented with the light in his factories, finding it to be superior in the drawing rooms and offices. After 1904, this initial success would fade, with competition from the tungsten lamp. The tungsten lamp had a major cost advantage over the aesthetically superior Nernst Lamp. Finally, Westinghouse would close the failed operation, but it had developed a new interest in the spectrum of light by Westinghouse engineers. It may be easy to look at some of these ancillary failures as systematic of a policy or personality fault, but that, would be a mistake. The finances are certainly open to criticism, but the aggressive technical development was at the heart of Westinghouse's successes. AC current had been a major risk as well as the turbo-generator. In both cases the more conservative management of General Electric suffered huge opportunity costs by passing initially on the technology. What is surprising, is not that Westinghouse had technological failures, but that there were so few of them. By 1899, Westinghouse had taken the advice of the Board of Directors, whose growing concern continued toward research and development dollars. Even after the end of the patent wars with General Electric by the patent control board, Westinghouse continued to purchase almost any invention that caught his fancy. Westinghouse Electric and Manufacturing Company formed a board of engineers to review commercial value. This forced Westinghouse to invest on his own as he had done in the past, but helped keep Westinghouse Electric more focused. Benjamin Lamme, Chief Engineer, would head this committee for years. This helped eliminate some of the small patent purchases by Westinghouse forcing a tough review by all concerned. The turbo-generator and electric railroads were big enough alone for Westinghouse Electric engineering resources.

In March of 1900, Westinghouse had purchased patent rights for the Cooper Hewitt Lamp, which was the forerunner of the fluorescent light. Passing a current through a gas illuminated these lights. The gas in this case was mercury vapor. Peter Cooper Hewitt had been working on the new light for almost six years. It offered new potential in the emerging film making industry as well as photo studios. The deal between Westinghouse and Cooper-Hewitt consisted of the formation of a company to exploit the new light. Westinghouse used his own funds to back Cooper Hewitt until a company could be formed. Once again, Westinghouse had to use personal funds to finance the company of Cooper Hewitt Electric Company in 1902. Westinghouse Electric's debt was just too high, even though it had strong cash flow and was paying a high dividend. The Cooper Hewitt lamp, though high in efficiency compared to incandescent lighting, had a harsh hue. The light lacked the red and violet light of the spectrum, and had too much green. Still, its efficiency drew Westinghouse into experimentation. Sales suffered because the lamp was cumbersome to use, containing over a pound of mercury in each.

During the early 1900s, Westinghouse also turned to the pursuit of electrifying railroads. Westinghouse looking at the overall efficiency of the steam locomotive, found it to be highly inefficient because of the need to transport and distribute coal for the locomotive. It was clearly more efficient to distribute electricity produced by coal-steam central electric stations. Westinghouse, with his experience in railroads, power generation, and electric motors, was in a unique position to evaluate electric railroads. Besides the coal savings, which were significant, speed and constancy of speed could both be improved as well as the size of the train. Westinghouse suggested over all maintenance labor would be reduced as well. Westinghouse Electric was already locked in a competition with General Electric for motors and streetcars, so this was a logical extension of Westinghouse's resources. General Electric had an early advantage with the merger of Thomson-Houston's streetcar technology with some of Edison's initial work. As early as the 1893 World's Fair, General Electric had demonstrated an electric locomotive for "yard" work at factories. General Electric gained critical engineering expertise with the Baltimore & Ohio Railroad tunnel in Baltimore. The Baltimore & Ohio was building a tunnel to save time on its New York to Washington line. The mile long tunnel had a significant upgrade and concerns for smoke and carbon monoxide in such a confined space required that an electric engine be used for the tunnel section. The first use of an electric locomotive for tunnels was the City and South London Tube Railway in 1890. The City and South hauled forty-ton trains at an average speed of 12 miles per hour. Westinghouse had actually ridden the City and London, and realized the huge potential, but General Electric (initial bid was pre-merger by Thomson-Houston) was the only company in 1894 that could manufacture the electric locomotive for the Baltimore & Ohio tunnel.

The Baltimore and Ohio tunnel locomotive of General Electric was a great success. It hauled forty-four freight cars up the grade at 12 miles per hour. A Baltimore & Ohio bulletin hailed the event with the following statement Not a sputter, spark, or slip of wheel indicated the tremendous energy which was developed by the locomotive." General Electric used its new GE-2000 heavy duty DC motor. Westinghouse believed for heavy electrified railroads, the AC single-phase motor offered a technological edge. Remember Westinghouse had used a rotary converter and DC motors for his lighter weight streetcars. The relatively short tunnel distances required no long distance distribution of electricity; therefore, direct current was economical. The same was true for short distance, lightweight streetcar systems. Westinghouse, however, had a bigger dream of electrified national railroad lines. This would require the transmission of high voltage AC current for long distances. The use directly of an AC motor in this application made sense, and could give Westinghouse a competitive advantage. The AC current would of course be needed to railroad long distances. AC motors also allowed a constant speed over long distances because the AC motor would handle voltage variation in long distance transmission. It was also a dream that Westinghouse had when he first purchased the AC motor patents of Tesla. The single phase AC motor has to rank up there with the rotary steam engine as a persistent dream of his. The development of the single-phase motor goes back to Tesla's first work at Westinghouse's Pittsburgh work. Tesla, Lamme, and Westinghouse had won a key technical advance in the use of 60 cycle AC coming down from 133 cycle, which allowed the AC motor to be used in factory applications. Tesla had long argued that higher horsepower was dependent on lower frequency.

After Tesla left in the early 1890s, Benjamin Lamme remained on this priority project of Westinghouse. By 1897 Lamme had developed a 60 cycle, single-phase AC motor for commercial applications. Westinghouse Electric started producing 5 to 40 horsepower AC motors of this type. The first big application came in 1897 at the Swift Company, in 40 horsepower cranes for their meat packaging plant. Another customer for these cranes was Baldwin Locomotive Works. Westinghouse hired George Gibbs to head the project. Westinghouse always liked to use project managers because of the need to coordinate all of the resources available in the Westinghouse companies. These project or matrix management approaches gave Westinghouse another advantage over General Electric, which tried to manage by division in the corporate structure. Gibbs was an expert mechanical engineer from Chicago, Milwaukee, and St. Paul Railroad with strong administrative skills. Unfortunately, to assure Gibbs's full commitment to the project, Westinghouse had to purchase his consulting firm. Here again we see Westinghouse showing a tendency to over spend for personnel or patents, but Westinghouse, in buying Gibbs Electric also brought another gifted executive into the Westinghouse organization. Edwin Herr was a Yale graduate with an international reputation in locomotive engineering. In 1911 Herr would take over the presidency of Westinghouse Electric, but initially Westinghouse started him at Air Brake with project consultation duties on electric locomotives.

The immediate issue, however, was a large locomotive. 40 horsepower was still too little power to drive a locomotive, but Westinghouse would not give up, and pushed Lamme and his team of engineers on to new goals. Lamme followed Tesla's earlier suggestion of lower frequencies. Earlier, as noted, Westinghouse had moved into a joint venture with Baldwin Locomotive Works to build electric locomotives. The key issue was motor size, but Westinghouse needed to strengthen his organization as well. He had Lamme working on the AC motor, but needed someone to coordinate the overall project. In 1902, Lamme commercialized a 100 horsepower single phase AC motor at a 16-cycle frequency. With the introduction of this 100 horsepower motor, its use in electric railroads became feasible. Westinghouse bid on a new tunnel application. It was in 1904 that initial bids were asked. The tunnel was the Grand Trunk Railway's St. Clair Tunnel under the St. Clair River. This cast iron tube tunnel would represent a major engineering challenge. Westinghouse felt this was the challenge he and his company needed, and he was willing to push the envelope. Westinghouse's effort for the contract focused on the International Railway Congress being held in Washington in May of 1905. Westinghouse lined up a special train to bring Congress members to East Pittsburgh to see two new systems. One was the friction draft gear, and a single phase, 25-cycle AC locomotive, which circled the East Pittsburgh plant. His engineers completed work on the locomotive only the night before, but its performance was perfect the next day, May 15. The locomotive moved fifty steel cars without the least bit of slipping.

The contract was let for a 25 cycle, single-phase system with a 300 horsepower engine. About the same time he also contracted for the use of a 300 horsepower engine on the New York, New Haven & Hartford Railroad. Westinghouse, many believed had this time gone too far. He had no 300 horsepower engine, and even some of his own engineers considered it a physical impossibility. General Electric made it known that they felt it could not be done. In the summer of 1905, Westinghouse called his engineers together and stated Now I have dropped you in the middle of a pond and it is up to you to swim out." Westinghouse was able to develop the 300 horsepower engine needed to the amazement of the engineering world. The St. Clair Tunnel ended in success, but the electrification New Haven project was too far ahead of its time. Some labeled it a "colossal failure," but in reality it would pull all of Westinghouse's resources into the electric train business. The New Haven line covered the distance between Woodlawn, New York, and Stamford, Connecticut. The New Haven system would ultimately take "the measure of the direct current, third rail system, eliminating its future use," the third rail, direct current system being that of General Electric. Another motivation for electric railroads came in 1903 and 1907. The New York legislature in 1903 passed a law to outlaw steam locomotives south of the Harlem River because of the dirt and soot around Grand Central Station. Westinghouse was now in a position to put his diversified technologies and companies into the overall broad requirements of an electric train system. George Gibbs would recall the accumulation of all these technologies in 1939;

Some 50 years ago, back in the 80s, when Westinghouse was evolving a new triple valve, developing an automatic signal system, testing the friction draft gear, investigating the alternating current transformer, and "playing" with his rotary engine, seemly scattering his efforts over many unrelated fields, no one dreamed, unless it was Westinghouse himself, that they would some day all converge in a single project and produce the premier railway electrification of the world.

Westinghouse also further promoted the use of steel in many of these elevated, tunnel, and subway electrification projects. George Gibbs had originally suggested the use of steel cars in tunnels and subways because of the danger of fire with wooden cars. Gibbs went on to suggest a simple design to Westinghouse. Westinghouse also gained inspiration from one of his old Air Brake directors, A. Cassatt, who was now President of the Pennsylvania Railroad. Cassatt was concerned about fires in the tunnels of the New York area. At the time all rail cars were made of wood. Cassatt offered to build cars at the Altoona Works of the Pennsylvania Railroad. The Rapid Transit Subway Construction Company purchased 300 of them. Westinghouse got the contract, but he had to deal with an old ghost in New York. The press started to talk of passengers being imprisoned in electrified steel 'coffins." August Belmont was the financer of the project, and helped assure its success for Westinghouse. While Westinghouse can be credited with the commercialization of steel cars, he lacked the capital to start his own company as the Panic of 1907 was on the horizon.

During the period of 1900 to 1907, Westinghouse Electric was moving into a new area of household appliances. The simple electric fan was one of the earliest of these appliances, which was a logical application of the AC motor. In 1886, Schuyler Wheeler invented the electric fan. Prior to that Philip Diehl had introduced the ceiling fan. The firm of Crocker & Curtis was producing fans by the 1890s. Early fans could use DC or AC motors, but by 1900, Westinghouse was producing a desk size motor for fans. In 1903, Westinghouse was producing a ceiling fan with its AC motors. Early fans required oil cups because of bearing leakage, again it was Westinghouse that improved to a self-lubricating bearing. Bearing development was one of the many unheralded improvements of Westinghouse companies that in aggregate revolutionized the electrical world. Electric fans followed by electric toasters and coffee makers became the earliest consumer products of Westinghouse Electric Research. These patents were a large group of corporate patents attributed to the Westinghouse organization. At his death Westinghouse and corporations would have over 15,000 patents under their control.

Chapter 16. The Old Genius and His Employees

> "The person who takes charity thinks himself inferior. The donor feels superior. I would rather give a man a chance to earn a dollar than give him five and make him feel he's a charity case."
>
> — *George Westinghouse*

The most pressing problem in the industrial towns of Pennsylvania was disability and accidental death. A survey of Allegheny county (Pittsburgh area) from July 1906 to June 1907 shows the magnitude of the problem. The total fatalities in this one-year period for the county were 526.[95] Disabling injuries were four times more. Prior to 1903, all of these workers and their families received no compensation for industrial accidents. Early efforts to help the worker came from the Roman Catholic Church and ethnic lodges, which offered some insurance packages. Westinghouse had personally done much to help such families since the beginning of the company, but the demands required a more formal approach. Westinghouse demonstrated the same creativity for these social programs. His benefit programs were so efficient that they command a new look for application in today's globalized competition. These programs were not altruism or socialism, but Christian based capitalism. The plans were always pragmatic and profitable with better benefits than those required by government. Westinghouse was generous but resisted the pure charity of other industrialists; he preferred to spend

95 Howard Harris, editor, *Keystone of Democracy; A History of Pennsylvania Workers* (Harrisburg: Pennsylvania Historical and Museum Commission, 1999), 129

money on personal improvement. He actually deplored straight, indiscriminate charity. He viewed himself as an industrial leader, not a trustee of the community's money.

Westinghouse pioneered mutual insurance and disability insurance plans. As early as 1884, Westinghouse Machine formed the Westinghouse Machine Company Mutual Aid Society. The company funded this separate organization in large part. Records of the initial plan could not be found but by 1900 there are some statistics. Membership was voluntary for the employee requiring $.50 a month. In the case of disability due to sickness the worker was entitled to $5.00 a week for six months, and $7.00 a week for disability due to injury. The death benefit was $100 from natural causes and $150 from accidental death. The company contributed one-third of the injury benefit and one-third of the accidental death benefit.[96]

Welfare benefits were a pioneering concept of Westinghouse, but as always with Westinghouse, it was a mutual benefit. Westinghouse established the "Relief Department" at Air Brake to help employees in 1903. The *Wilmerding News* described the department mission in 1904 as:

> To insure a certain income to the employees who might become unfitted for work through illness or injury and in the event of death to pay the beneficiary a stipulated sum. Any employee under 50 years of age is entitled to membership, subject to successful physical examination, but membership is not compulsory. Members contribute according to the class in which they belong, there being five, the class being determined by the wages received, varying from $35 to $95 or over per month, the contribution ranging from 50 cents to $1.50[per week]. A member may receive benefits for 39 consecutive weeks, in event of disability extending so long a period. The air brake company is the custodian of the funds, but being such does not benefit the company pecuniarily, as it pays four per cent interest on monthly balances to the credit of the relief fund. The company goes further and guarantees payment of all benefits, and if the money received from the monthly contributions be insufficient to meet the requirements the company makes good such deficit.[97]

Westinghouse companies were self-insurers for mutual benefits using the company financial resources as collateral. The company and the program were independent, except that the company bore the expense of administration. Participation in the program was voluntary; the disability benefits covered about three-fourths of the Westinghouse employees. The rates for the life and disability insurance offered were somewhat lower than those of private insurance companies (.70 to $1.70 per week) of the time such as Prudential and Home Guards. The interest paid on benefit investment of 4% was unique. The weekly payment still represented a significant portion of the weekly wage, but most employees invested. In addition, the Westinghouse plan was an open plan in which steel workers in hazardous jobs who

96 Crystal Eastman, *Work Accidents and the Law* (New York: Charities Publication, 1910), 158

97 *Wilmerding News*, November 23, 1904

previously could either not get private insurance or had to pay increased rates. Of course, Westinghouse plants were safety oriented, and had a significantly lower accident rate than the steel mills of the area. In addition, Westinghouse plants had doctors and nurses available to employees daily at the plant. Westinghouse employees that were disabled to a lesser degree were often given light work assignments to keep them employed. Westinghouse programs were never socialistic, but remained capitalistic in structure. In many ways, the Westinghouse programs fostered a competition for the best labor in the Pittsburgh area. Eventually, in 1908, even the Morgan controlled Carnegie Steel was forced to offer some disability insurance.

By 1913,two years before the Workmen's Compensation Act in Pennsylvania, Air Brake started to fully fund a Workmen's compensation program. The fund was totally maintained by Air Brake with no payments by the worker, and all workers were covered. Again, Westinghouse pioneered the way for what would become state law. Even with the law, Air Brake payments were much more generous, and payment began immediately while the state allowed for a ten day waiting period. The Westinghouse model was the basis for the design of the state system. Westinghouse's early interests in such benefit programs, not only stimulated benefit competition, but also gained the focused opposition of powerful people, such as J.P. Morgan and Henry Clay Frick. Westinghouse tended to attract the best employees even during times of labor shortages.

Westinghouse saw his Relief Department as a major success as well and extended it to both Air Brake and Westinghouse Electric. Similarly, Westinghouse had one of the first pension plans for his employees. There is much evidence that Westinghouse had an informal pension plan in the 1890s based on the discrimination of the board of trustees. The 1906 pension plan of the company is attributed to John Miller, then secretary of the company, and later its President. The plan required no employee payment and was based solely on years of service and was not at the discretion of management. Pensions did exist at Carnegie Steel in the area, but were based on a demonstration of need after "long and creditable service." The Carnegie Steel plan known as the Carnegie Relief Fund was actually a gift of Andrew Carnegie in his retirement to his old company, and was a trust of four million dollars to be administered by Carnegie Steel. Carnegie's boyhood friend and Westinghouse Air Brake director, Robert Pitcairn, was named by Carnegie as administrator. The Carnegie plan was not financed by the company, but functioned from Carnegie's gift.

The plan applied to Carnegie Steel (mainly in Pittsburgh), and not the larger entity of United States Steel. This limited Carnegie pension Plan went into effect in 1901.[98] In reality, it was Carnegie competing with Westinghouse for image, but more fundamentally, it was a difference in philosophy. Carnegie never believed in a company plan, but fostered the capitalist trust-

98 Margaret Byington, *Homestead: The Households of a Mill Town* (Pittsburgh: University of Pittsburgh Press, 1974), Appendix VI

ee approach. Specifically, Carnegie aimed at the most needy and "acceptable" making it charity. Certainly, this is not to take away from Carnegie, but it was not a true company pension plan. The Westinghouse plan was a true company plan, worked out by a professional actuary, and it was extremely liberal even by today's standards. One biographer suggested that the program was started by the removal of the holiday turkey give away, but that could have only been a small deposit on such an extensive program.[99] This program allowed an employee to retire at seventy to sixty-five years of age depending on circumstances, having at least twenty years of service. Even more amazing was that the pension benefits continued to his dependents after death. The fund was maintained in a separate account that insured the funds even in the case of Air Brake going out of business, an idea far ahead of its time.

While it was an inherent belief that it paid to treat employees right, it went much deeper than that. Westinghouse was a deeply moral person and had a genuine concern for people, and this belief is clearly traceable to his New England Puritan upbringing. Certainly, his highly profitable company allowed extremes of employee compensation that maybe others could not. Westinghouse's first biographer, Francis Leupp, reports that in the early days of the Air Brake pension program, a number of loyal employees had reached the age of seventy but were a few days short of the required twenty years of service needed for a pension. When Westinghouse heard of the situation, he forced the board of directors to adjust the plan. This was far from an isolated case. Westinghouse many times overruled his managers in keeping employees through bad times. Westinghouse's approach went back to his family roots. Unlike Andrew Carnegie, Robert Owen, and so many other charitable industrialists, he never wrote down or defined his approach. His behavior defined his model.

At Union Switch and Signal, a few years later, Westinghouse pioneered an employee stock program. This experiment was visionary, but it was consistent with Westinghouse's idea of employee ownership. The plan allowed employees to buy stock, but was not a matching program. This was a time when the mechanics of the stock market and capital requirements limited the ability of the worker to buy even a small amount of stock. The plan allowed for small installments and gave a small bonus after five years. The money would be taken directly out of the paycheck; or the local savings & loans were brought into the purchasing program. This type of automatic plan was at least fifty years ahead of its time. The employee stock plan overall was one of the example in which Westinghouse followed his neighboring industrial giant, Carnegie Steel. Carnegie had promoted stock ownership with his management as early as 1878. The Carnegie employee stock plan saw little success because of the lower pay of the steelworkers. Stock ownership was also one of the few employee programs that bankers such as J.P. Morgan

99 Leupp, 248

found acceptable. In fact, it was George Perkins, a Morgan partner on the Carnegie Steel board, who devised one of the first stock programs in 1901.

Actually, the motivation for Carnegie Steel and Westinghouse to implement an employee stock plan was the same. The collective wisdom at the time was that it offered an effective deterrent to unionization. The idea was that ownership would replace union protection, as well; and the loss of dividends due to strikes would make workers hesitant to strike. In theory, it made sense, but even today the amount of stock ownership available rarely translates into a feeling of ownership. The Westinghouse plan did fare better than that of Carnegie Steel, but it was difficult for average workmen to add stock purchases to their budgets. Carnegie Steel laborers mostly found themselves lacking the money to participate. The plans were popular with professional employees and management at both companies.

The real difference in the Westinghouse approach to his workers can best be seen in the working conditions and plant environment. As noted, Westinghouse employees in the 1880s have been assigned a five-and-a-half day workweek with Sundays off, while Carnegie employees worked a seven-day week with 12-hour shifts into the 1920s. Westinghouse plants, as early as 1890, had washrooms and toilets, something lacking in the great steel mills of the Pittsburgh area and most of the factories across the nation. Lighting and ventilation were part of all Westinghouse factories. A 1904 Westinghouse Electric promotional book described his East Pittsburgh Works as follows:

> Much care has been given to the sanitary condition of the shop, and the comfort of the employees has been carefully considered. In the offices and Works an even temperature is maintained by means of the most improved ventilating systems. The shop is lighted by Bremer arc lamps and Nernst lamps; the offices by Nernst and Sawyer-Man incandescent lamps. The Cooper-Hewitt Mercury Vapor lamp is also used for lighting the drawing offices. Several artesian wells in the shop furnish drinking water for the employees. Well-kept wash rooms, coat lockers, and toilet rooms are distributed at convenient points throughout the Works. On the sixth floor of the office building the company has provided a retiring room and a lunchroom for the lady employees.

Westinghouse Electric had become the largest employer of women in the United States. In 1910, Westinghouse Electric had about 10,000 employees, of which 1,000 were women. Even in his earliest lunchrooms, Westinghouse required special tablecloths for the women employees. Special walkways were designed for women employees as well to assure that their dresses were kept clean. The women's lunchroom at Westinghouse Electric had thirty-five tables and could accommodate over a thousand women. The company provided for hot coffee. Two maidservants were on duty to carry dishes, pour coffee, etc. The cleanliness of Westinghouse plants would rival the Japanese plants of today. The stockholders and bankers of Westinghouse Electric often complained about the costs of many of these plant improvements. Westinghouse actually was a pioneer in treating women fairly in the workplace. He believed that workingwomen required the same dignity as the non-working Victorian ladies of the time. He demanded that his male employees be proper at all times.

Westinghouse's views of women were Victorian, but Christian and caring. He realized the need for women to work to help the family, but he was concerned about their future. He believed that factory assembly work for women should be a temporary career. Most immigrant families found it necessary for wives to work for a few years in order to build savings for housing. He developed training schools for women employees. These schools operated in the afternoon and evenings offering courses in "stenography, typewriting, cooking, sewing, household art, and music." The turnover of women in Westinghouse plants was generally very high, as women moved on to their families or office jobs after two to four years, but women were among the most loyal employees in the organization because of Westinghouse's fatherly concern for them.

Westinghouse had a similar approach to the young boys he employed. He offered free education in the trades, and college scholarships where talent merited it. This, of course, was in addition to the programs offered by the Y.M.C.A. In 1888, Westinghouse started a type of scholarship program that recruited promising students offering them education and a job after graduation. At his East Pittsburgh Westinghouse Electric Works, he provided bowling alleys, poolrooms, and reading rooms for his young employees. He purchased land around the Pittsburgh area to function as weekend camps for the boys as well as for the Y.M.C.A. In addition, he developed several employee family parks in the Pittsburgh area for his employees. Westinghouse also sponsored a number of clubs for his employees, such as the "Electric Club" and the "Amber Club," which fostered the study of technology. Westinghouse built a meeting and headquarters office for the Electric Club, which also offered rooms for single engineers in transit. Westinghouse himself would often attend the Sunday afternoon meetings with these young employees. These clubs published technical newsletters and promoted interesting projects. The club on a weekly basis offered various popular lectures. The "Electric Club" could list four members that became president of the American Institute of Electrical Engineers, as well as great scientists, such as Benjamin Garver Lamme and Bertha Lamme. Bertha Lamme, the sister of Benjamin, is today recognized as the first professional woman electrical engineer. Westinghouse would often sit with Bertha on their train commute from Wilmerding/East Pittsburgh to the Pittsburgh suburbs, encouraging her.

Westinghouse loved to visit his younger employees on the job as well as in the clubs. While he nurtured the technical skills of these young employees, he also passed on an ethical and moral code. Henry Prout noted in 1921, this impact There soaked into the mind of those young men, unconsciously, not only lasting contempt for what was off color but deep disdain for cunning and craft, and for dishonesty, moral or intellectual." Prout further noted that, "They learned, too, to be careful not to boast."[100] Westinghouse was a man of deep character that defined a culture of character in all his organizations. Westinghouse had a true hatred of personal publicity feeling it to be a sin of pride; again we see the roots of his mother's religious views. Unlike Carnegie, many of Pittsburghers of the time knew little of Westinghouse. He stayed away from newspapers and kept all charitable donations anonymous.

The Westinghouse system was a powerful application of New England machine shop management in the new industrial revolution. Westinghouse pioneered the electrical apprentice program in the United States. Both in machine apprenticeships and electrical apprenticeships, few companies approached the numbers of Westinghouse Electric and Air Brake. He created an ordinary apprentice program for non-engineering grads and an engineering program for graduates. Preference to enter these programs was always given to employees and family. Westinghouse was soon generating more apprentices than he needed, but the demand was high in the Pittsburgh area. Bankers like Morgan saw the high turnover as wasteful. Westinghouse argued successfully that engineers and apprentices that left for other companies performed a valuable function through their ability to install its products throughout the world. Westinghouse's final struggle with the Morgan management was to keep the employee programs going. Morgan ultimately gave in realizing culture would be costly to the organization and he would be fighting the very roots of a management indoctrinated by the Westinghouse style. Westinghouse Electric would even today maintain part of the original culture George Westinghouse imbued it with.

While Westinghouse had revolutionized employee and human relations management, in retrospect, one group might feel he didn't go far enough. Westinghouse Electric at the East Pittsburgh plant was a major employer of women in the Pittsburgh area. While Westinghouse had gone farther than any employer in treating them with dignity and respect, he followed most of the norms of the day. His women were paid about $9 a week in 1907, which was far above the $6-7 average in Pittsburgh. Still, men in Westinghouse plants made double that, which was standard. Westinghouse jobs were still highly prized by both men and women. The East Pittsburgh plant of 10,000 employees had 12,000 applications a year. Sales jobs for women were considered more dignified but were filled by the native groups of Scotch-Irish, Irish, and German, not the more common labor pool of immigrants. Italian women usually would not work because of family beliefs. The Poles, Slavs, and Hungarians flocked to the factories. The jobs at East Pittsburgh paid well but they were hard work. Massive amounts of women were used to wind motors, build motor cores, and apply insulation. Women also performed rework because of their cheaper rate. Piece rate was the usual method. With overtime,

100 Henry Prout, *The Life of George Westinghouse*,(New York: Cosimo Books, 1921)

a woman at Westinghouse in 1907 might make $12 a week. There was a high turnover rate but the reason for this remains unclear. Marriage, return to the family, and pregnancy probably were bigger reasons than the mental strain suggested by some. One reporter claimed that few women worked more than four years.[101] There were complaints of rate cutting when women started to achieve good production levels. While Westinghouse seemed to believe in a higher standard, his foremen were less enlightened.

Many have pointed out that Westinghouse often allowed women to operate machinery, something, not common in other industries of the time. Still, there were business advantages to this practice. The women replaced men at half the cost. Studies of women in other industries in 1907 Pittsburgh show that Westinghouse at least had improved working conditions. Lighting and ventilation were particularly poor at neighboring industries. J. H. Heinz hired a large number of women in the area as well. Heinz women had a similar profile being young, 16 to 20, and primarily Slav and Hungarian. Heinz records showed that once they married they rarely returned to work. Top pay in the canning industry for women was around $7 a day. Another industry many women went to was the tobacco or stogy rolling factories. These were known as "sweatshops," with low pay and poor ventilation and unsanitary conditions. We have already seen that in his first Air Brake operation in the 1870s, Westinghouse instituted the half-day Saturday, and Sunday off. While much has been made of this early Sunday off practice, he was not the first to offer it. Prior to 1870 and the industrialization of Pittsburgh, Sunday was always a worker day of rest. The Presbyterian core of the city had reinforced Pittsburgh's early Sabbatarianism. By 1875, with the rise of the steel, railroad, and glass industry, all had changed. The Scotch-Irish masters of steel and glass did not feel their Sabbatarianism applied to the Southern, Catholic immigrants in the workforce but not with the Westinghouse plants. Massive steel plants needed to run on a 7 day, 24-hour schedule. Westinghouse went even further at his East Pittsburgh plants by implementing the nine-hour day. For today's reader, this may hardly seem humane, but the standard of the day was the 12-hour day. The great neighboring steel mills of Carnegie, worked a 12-hour, 7-day week. There were no holidays. It was a brutal schedule. The 12-hour day was common in heavy industry until the 1920s.

Westinghouse observed much in the progressive British industries in his many trips in the 1870s. One company that had a powerful impact was the Soho foundry. The Soho foundry of Great Britain was the engineering company of James Watt and Mathew Boulton, originally formed in 1800 to build steam engines. Soho foundry had implemented piece rate production fifty years before anyone. The results had shown higher production and lower unit costs. Westinghouse was keen to observe that this was heavy industrial production that looked like the management of New England machine shops.

101 Elizabeth Beardsley, *Women and the Trades, Pittsburgh, 1907-1908* (New York: Charities Publication Committee, 1909)

The Soho foundry was clearly founded on Christian principles in which the employee had value beyond machines. The religious views of workplace were consistent with Westinghouse's New England background as typified by this Boulton quote As the smith cannot do without his striker, so neither can the Master do without his Workmen. Let each perform his part well and do their duty in that state which it hath pleased God to call them, and this they will find to be the true rational ground of equality."[102] Soho professed that worker programs were a necessary requirement to maintain the best work force. Where Westinghouse would ultimately differ from Boulton was that while paternal, he believed in the independence of the worker from the master in belief and social norms. Westinghouse never in anyway tried to influence the religious beliefs of anyone, except by his personal behavior as a Christian.

Still, there was much to learn and emulate about Soho. The foundry walls were whitewashed and kept extremely clean, something rarely seen in industrial Britain or America. Cleanness was demanded at Soho in all aspects, because of the respect for the dignity of the worker. Westinghouse demanded a high standard of cleanness in all of his factory washrooms. The Soho also was known for giving Christmas gifts and raises at Christmas time. Westinghouse would become just as famous in America for his Thanksgiving turkeys and family dinners. Soho rented company homes to its workers. Westinghouse preferred employee home ownership over the company rented house. Probably, the most progressive Soho benefit was its Mutual Assurance Society, the first company insurance plan of the modern era. It is unclear how many of these ideas Westinghouse borrowed; they were all consistent with his New England manufacturing experience. Another necessary function that Westinghouse similarly adopted was the use of cost accounting. Cost accounting was necessary to calculate and understand unit costs. Unit costs are needed to evaluate piece rate and incentive plans. All of these applications also offer further proof of the superiority of Westinghouse as a businessman. While great paternal business leaders such as Carnegie saw the power and magic of incentive plans for his managers, they lacked the vision to take it to the average worker. Westinghouse did not segregate men by their position, as Carnegie. Furthermore, Westinghouse never saw his employees as children or childlike, and that differentiates him from Boulton, Watt, Carnegie and many others, although, in fairness, that paternal treatment is preferable to looking at workers as just another input, as many bankers of the time such as Morgan did.

Westinghouse's concept of work is often a contribution that is overlooked. He freely mixed in the "Golden Rule" without preaching any religion. Westinghouse applied his basic and orthodox Christian beliefs in the workplace without encroaching on individual beliefs. His concept was very Benedictine in that he viewed work as part of a natural religion. Men were

102 Erick Roll, *An Early Experiment in Industrial Organization* (London: Longmans, Green & Co., 1930), 222

born to improve the world through work. He preferred to give the gift of work and improve the life of work rather than practice the more traditional means of charity. He was always concerned about his employees being laid off because of its mental impact. Probably Walter Houghton best explains this Victorian concept of work of Westinghouse:

> The Victorian gospel of work, derived from its religious and economic life and preached the more earnestly because the idea of crisis and the idea of progress both called for dedicated action, found further support from an unexpected quarter. As the difficulties of belief increased, the essence of religion for Christians — and for agnostics the "meaning of life" — came more and more to lie in strenuous labor for the good of society. That was not only a rational alternative to fruitless speculation but also a means of exorcizing the mood of ennui and despair which so often accompanied the loss of faith. For these reasons, a religion of work, with or without a supernatural context, came to be, in fact, the actual faith of many Victorians: it could resolve both intellectual perplexity and psychological depression.[103]

To a large degree, the measure of George Westinghouse could be found in his vision of ethics, work, and fair play that became embodied in the mission and vision of Westinghouse Electric and his other companies. Westinghouse Electric's Charles Scott, executive and famous engineer, described it best in his summary of his Westinghouse experience:

> In my thirty-five years of work with the Westinghouse Company, I have seen many young men grow from pupils to assistants and associates. This has been one of my greatest pleasures. I have aimed to instill in them fundamental ideas of engineering, honesty, and honor, square dealing and fair fighting —that there should be pride in accomplishment because true engineering means much more than merely making a living; it means advancement of the art for the benefit of mankind.[104]

It was the kind of career experience that Westinghouse would have wanted from one of his pupils and employees. Westinghouse imparted a missionary zeal that has remained with Westinghouse Electric for over a hundred years. With the exceptions of Henry Ford, Steve Jobs, and a handful of others, few company founders have left such a defining mark on corporate culture.

Westinghouse's management style as noted earlier was paternal, but it was also structurally hierarchical and decentralized. Westinghouse loved to drop in and discuss things that were parts of his current objectives, but it would be a mistake to visualize this as a father out patting his sons on the back. Though he could be a bit terse and preoccupied in person on his plant visits this probably related to the intensity of thought and attention he put into his projects. Many times he would pass his workers with only a kind smile. He tended to follow his pet projects with a passion while overlooking others who were jealous of the attention. Jill Jonnes, author of *Empires of Light*, recounts the experience of a boy employee who had dropped a slab of copper. While others laughed, Westinghouse nearby in his classic long coat,

103 Walter Houghton, *The Victorian Frame of Mind* (New Haven: Yale Press, 1959)
104 Lamme, introductory page

gloves, and top hat came over to help the boy lift the slab without saying anything. It was a silent act of kindness that the boy recalled almost fifty years later. He was not a managerial cheerleader, but an organizational motivator. Rewards and encouragement came through organizational channels. Of course, things were different at his dinner parties, although Marguerite seemed to have added the social lubricant to those parties. If you did get to Westinghouse's office or ear, he would get involved but he preferred to have a just organization that didn't require him to act as a judge. He allowed his managers great leeway to motivate and assure justice in the work place. He had a big heart and cared deeply about his employees and he expected his managers to reflect his big heart. He wanted a caring organization that would be his legacy, rather than one dependent on his presence. Like most of the Gilded Age capitalists, with the exception of Thomas Edison, he believed in a Napoleonic decentralized operation. Again his pet projects were always problematic to the organization. Sometimes his zeal would lead him to disregard organizational structure. He was always an engineer at heart with a love for experimentation. Still, even in these cases, he respected his managers' opinions.

When a lieutenant was asked during a conference why managers stayed loyal to Westinghouse, the answer was:

> No more than the marshals of Napoleon yearned to set up shop for themselves. Why should I? Why should any of us? Mr. Westinghouse is fair, liberal, and just. He has the spirit of a really great empire builder. He has supreme courage. He lacks entirely the conservatism that leads to content and dry rot. We are always doing something new, always going forward. We are never satisfied. It has been aptly said that 'Mr. Westinghouse is always in competition with him self.' He always subordinates so-called 'business considerations' to engineering perfection, so much so that all of his enemies and some of his friends have charged that he is on the whole a poor business man. We believe this to be the very foundation rock on which his success has been built.

It almost seems unbelievable that the charge of being a poor businessman could stick. By now it is surely it clear that the charge was a fabrication of J. P. Morgan and associates who controlled the dominant New York press. Westinghouse, however, had formed over 40 successful companies some of which still exist today. He had over and over again beaten the trusts and larger competition. No one, not even Vanderbilt or Carnegie, had avoided the grasp of J. P. Morgan and the trust banks, and even then only Westinghouse Electric was subdued. Westinghouse envisioned capitalism at its best. He had ultimately shown that cooperation was possible while still maintaining competitiveness.

The efficiencies reached at the various Westinghouse companies revealed a winning combination of people and innovation. Production at the Air Brake plant at the end of the 19th century reached a phenomenal height, a complete set of air brakes every minute, valued at $40 for freight cars and

$100 for passenger cars.[105] The partnership with Baldwin Locomotive led to an interest in the use of the procedures of Fredrick Taylor in efficiency engineering. Actually, Westinghouse Air Brake was one of the first companies to implement the revolutionary Taylor practices. Fredrick Winslow Taylor was born in 1856 in Philadelphia. Working at Midvale Steel in the 1880s, he developed the principles of scientific management. These practices focused on studying work and breaking it into smaller controllable elements. The sequence of work was then standardized. The Taylor system improved efficiencies by 50 to 60% through the application of better management. Westinghouse Electric became even more interested in Taylor's work when one of his studies showed that the main inefficiencies in belt-driven factories came from the belt themselves. Westinghouse electric motors could eliminate belt driven machines, which brought them into scientific management. Westinghouse would adopt the principles of scientific management at all his plants and be on the forefront of this revolution. Fredrick Taylor would become known as the "father of scientific management." His legacy is somewhat mixed, but it is clear that his ideas were extremely successful in the Gilded Age. Interestingly, one of the earliest opponents of scientific management was J. P. Morgan! In fact, when Westinghouse initially started to use Taylor principles in the 1890s, Taylor was known mainly in the machining industry. In 1910 Morgan lost a major lawsuit, which prevented a railroad rate increase because of his not applying Taylor principles. That lawsuit in 1910 made Taylor a "celebrity."[106]

There have been some misconceptions at the application of Taylorism at East Pittsburgh resulted in the 1914 strike there. While this strike can be related to piece rate standards, it does not fully reflect Westinghouse or Taylor's concepts. Westinghouse at Air Brake had used the piece rate system in the 1880s before Taylor's system. Westinghouse at East Pittsburgh using scientific management adapted the Westinghouse piece rate system. Westinghouse's approach was unique in that employee committees were involved in setting the piece rate versus pure management. It in hindsight was a novel and effective application of a piece rate pay system. George Westinghouse had anticipated the problems of piece rate pay if not fully set. After Westinghouse left active management in 1911, the plant moved to management set piece rate standards.

While Westinghouse was an ideal for engineers, he defined engineering management. At the dedication of the Schenley Park Westinghouse Memorial in 1936, the then President of the American Society of Engineers said this:

> Not merely an inventor or technical expert, he was an engineer in the broadest sense. To make Nature's elements of use to mankind involves the recognition of a need or an opportunity, the invention of some new method

105 Edward Williams, "Effect of Invention upon Labor and Morals," *The Chautauquan*, March, 1899

106 Morris, 307

> or device, and the development of the idea into concrete and practical form; then preparing for its production, convincing others of its usefulness, and supplying it for service. All these things Westinghouse did. He had facility in dealing with men and organizing them for research and development, and in manufacturing a facility in dealing with physical things, a rare combination of complementary qualities of the two men who made the firm of Bolton and Watt successful in launching the steam engine.

It might even be said that no other figure in the electrical field had this rare combination. Edison, Thomson, Tesla, and others lacked it. Westinghouse was able to delegate readily, except when it came to his "pet" projects, such as the steam turbine. Still, even with his pet projects where he showed intensity and passion, he was willing to listen and change at the suggestion of his engineers. He often had his engineers to dinner at Solitude to assure informal feedback on projects. Westinghouse seemed always to drive towards bigger goals, and saved his persistence, determination, and perceived hardheadness for those major goals. In his quest for electrical power generation, he showed flexibility on almost everything, including DC versus AC. Edison, on the other hand, could become stuck on a device or system. Westinghouse was always willing to incorporate others' ideas through purchasing patents, and he was always willing to give credit.

Paul Cravath, friend, partner, and patent lawyer, marveled at Westinghouse commitment for the end goal and ultimate triumph:

> I saw him thus intimately under almost every conceivable condition-in his home, at his office, in his factory, in his private car which was almost another home, abroad, as well as in this country. I saw him when he was elated by successful achievement, amid disappointments and discouragement, and more than once in the face of threatening disaster. I saw him when he was carrying a load of responsibility under which any other man whom I have ever known would have fallen. He was always the same; simple, unassuming, direct, frank, courageous, unfaltering in his faith, and supremely confident in the ultimate triumph of his plans. I have seen him wearied almost beyond endurance; disappointed beyond expression over some miscarriage of his plans; wounded in his feeling because he had discovered stupidity where he expected intelligence, discouragement where he expected encouragement, disloyalty where he had a right to expect loyalty. I had seen him more than once when every man about him despaired of his being able to attain the ends for which he was striving and advised surrender or compromise, but I have never known him to acknowledge defeat nor to yield to discouragement, nor to falter in his efforts to accomplish his main objectives.

Certainly, this is powerful testimony to the project bulldog that Westinghouse was.

In 1904 the working conditions and pioneering efforts of the Westinghouse companies were highlighted at the Louisiana Purchase Exposition held at St. Louis (St. Louis World Fair). These films were some of the first documentary films (known then as "actuality" films) made. The films represent the first motion pictures of American factories, and 21 of the original

29 are still available at the Library of Congress. American Mutoscope and Biograph Company produced them in the spring of 1904. These films covered Westinghouse Electric and Manufacturing at East Pittsburgh, Westinghouse Machine Company at East Pittsburgh, Westinghouse Air Brake at Wilmerding, the Westinghouse foundry at Trafford, and Union Switch and Signal at Swissvale. The films were made using the Cooper Hewitt lamps manufactured by the Westinghouse owned Cooper Hewitt Electric Company of New York. Without these lamps filming inside the factories would have been impossible. The films feature a unique study of women in the workplace doing, such tasks as assembling generators, mica slitting, testing motors, and winding coils. The films go over industrial procedures and Taylor efficiency practices, as well as, safety precautions. Prior to the films' release at St. Louis, Westinghouse planned a night viewing at Pittsburgh's Carnegie music hall.

CHAPTER 17. HIS LAST CHAPTER

"George Westinghouse is a man that can't be downed."

— Andrew Carnegie

Personally, the years 1904 and 1905 marked major events in Westinghouse's life. One might have expected Westinghouse to embrace the emerging automobile, but Westinghouse was a railroad man. He had learned to travel comfortably by rail. His private station and railroad car made distance travel enjoyable, as he traveled between New York and Pittsburgh offices and a trio of homes in Pittsburgh, New England, and Washington. He had refused to purchase the new automobile, until in 1904, his French Works built him an elaborately equipped limousine as an example of their craftsmanship. Marguerite took to its use first, and over the years Westinghouse started to use it more and more. It allowed him to commute quickly between Solitude, Pittsburgh, and the Turtle Creek Valley. He immediately hired a chauffer so Mrs. Westinghouse could use it for shopping trips to Pittsburgh, and New York when in Lenox. Westinghouse electric motors were also common in electric cars of the time, but Westinghouse did not exploit this market.

In 1905 Westinghouse was asked to serve as a public servant in a national securities issue. A scandal had broken out in the Equitable Life Assurance Society, which was a national life issuance stock company. It was one of the largest in the nation with $400 million in assets and 6 million policyholders. Its board of directors was a collection of America's greatest capitalists including Westinghouse's Pittsburgh neighbor, Henry Clay Frick. The Equitable scandal would rock Wall Street and Washington, and put fear into millions of small policyholders. The great scandal commanded 115 front-page articles in the *New York Times* and 122 in *The World* in a single year.[107] The scandal would tarnish New York's highest society and disturb the normalcy of business in Morgan's library. It would all start of the eve of January 31, 1905 with one of New York's greatest balls of the Gilded Age. It was given by James Hyde, president of Equitable, at the Sherry's Hotel, which had several floors converted into a Versailles type palace. The Sherry Hotel at the time was the most exclusive in New York, being the temporary home of Henry Clay Frick since 1902, with a museum quality art collection. The ball included many of the Equitable directors, such as Astor, Frick, Vanderbilt, Belmont, Gould, Cassatt and Winthrop. The ball itself made headlines, but its rumored cost of $200,000 ($4 million in today's dollars) shocked the sensibilities of many. Rumors started that Equitable had footed the bill. The gossip and rumors would start a panic that opened the company up to a charge of financial misuse of funds. While the link to the ball was never proven, funds had been used to pay fake consultants and directors, buy political favors, and numerous types of fraud. Frick, who had no personal stock ties was selected as chairman of an internal committee to correct the issue, which was causing a panic on the stock market. Frick could not get reform and resigned from the board. August Belmont and Andrew Cassatt also resigned. Many believed that J. P. Morgan and his rival insurance company New York Life were waiting to take over forming an insurance trust. A congressional investigation opened up many political ties as the scandal spun out of control. The stock drop had taken Equitable to the edge of bankruptcy and the politicians both nationally and in the state of New York needed a way out.

A New York financier stepped in purchasing a majority position and setting up an outside committee to reorganize Equitable. He needed and selected men that were beyond reproach to quickly rebuild public confidence. A troika of former President Grover Cleveland, Morgan O'Brien, a presiding Appellate judge and former Supreme Court Justice, and George Westinghouse. Westinghouse had the only organizational management experience, but it was a major strength. Westinghouse was considered by some a flawed candidate, since James Hyde's father, Henry had been a director on the board of Westinghouse Electric. While James had technically inherited the director seat, he never used it. Westinghouse and Henry had gone back to 1885, when a Westinghouse steam engine powered an elevator for the Equitable Building. The elevator allowed the Equitable Building to be the largest in the world at seven stories. Westinghouse, already overloaded with his companies, needed convincing by his friend and New York lawyer, Paul Cravath. Once Westinghouse accepted, the policy owners received ballots, and returned total support to the committee of three. Ryan transferred stock into a trust to assure the three could act independently and free from any financiers. The troika was given unrestricted authority to reorganize the

107 Patricia Beard, *After the Ball* (New York: Perennial, 2003), 5

company. They set out to bring new and competent managers. Within a year they reformed the company and headed off what had been called the "Rich Men's Panic." Westinghouse was showered with praise as Morgan fumed at Westinghouse's success. Another interesting note of the scandal was a drawing closer of Morgan, Frick, and Belmont, which would be a future problem for Westinghouse. The *New York Sun* hailed Westinghouse as the "world's greatest engineer," "an architect of America's Golden Age," and a "truly fine, kindly and lovable human being." Westinghouse was probably the only major capitalist of the time that was not tarnished by the scandal, but whose image improved from it. Unfortunately, it was time that cost him dearly as financial imbalance continued to be a problem for Westinghouse Electric and Manufacturing. Thanks to men like Herman Westinghouse at Westinghouse Machine and Edwin Herr at Air Brake, these companies were in outstanding financial shape.

With the loss of Westinghouse Electric in 1907, George Westinghouse was never quite the same. Work became tinged with a personal struggle with depression. The loss hurt in very strange ways. Westinghouse Electric had become the focus of his creativity. He had left the management of the other companies, such as Air Brake to others. Westinghouse Electric had become his hobby and outlet for his creativity. Loss of control hurt him deeply, but Marguerite played a key role in helping him fight on as President. Some of the hurt came from the loss to Morgan personally, and some from the many friends that deserted him. The damage to his image might well have been the most hurtful of all. The press attacks on Westinghouse were unfair and clearly the result of Morgan trying to put a spin on the take over. They were extremely tough to endure because they served little purpose other than to frame Morgan and the bankers as saviors. In this respect Pittsburgh let him down as a city. The Chicago press had taken on Morgan and his trusts in 1893, but the Pittsburgh press always followed New York and the *New York Times*, which had strong Morgan ties. Morgan was facing government scrutiny not only for the Westinghouse "reorganization," but the merger of Tennessee Coal & Iron into United States Steel. Tennessee Coal & Iron was a Dow Jones average stock at the time. The Pittsburgh press tended to support US Steel, and thus supported Morgan. Unfortunately, Westinghouse Electric took a back seat to Tennessee Coal & Iron because of it making United States Steel a horizontal monopoly. Horizontal monopolies, with their control of raw materials and production, appeared to be the most feared versus vertical trusts in specific markets. Westinghouse did not formally consist of a monopoly with General Electric, but it certainly led to some control by a Morgan-like trust. In these last years of his life, he spent most of his time away from Pittsburgh. His winter home was Blaine House in Washington and Erskine in the summer.

Morgan enlisted his stable of industrialists to drop statements to the press. As noted, even Carnegie added to the choir. Pittsburgher and President of United States Steel, Charles Schwab, held to the same line in a discussion with business reporter Clarence Barron, "George Westinghouse lived and slept airbrake, but when it came to other things he could not give them his attention in detail. Had he devoted himself the same way to Westinghouse Electric he would have made the same success, but a man cannot follow the details of many things."[108] This is totally off the mark, and clearly meant to please Morgan. Westinghouse Electric was George Westinghouse, in all aspects. Morgan surrogates continued to build an image of poor management, which one is hard pressed to find evidence of. The force of Morgan's spin, however, remains even today in many studies of the times. Again, what are found are top industrialists (all with Morgan connections) making insinuations. Specifics are always lacking. If any validity can be found in these claims, it would be in Westinghouse's financial management. Today such debt versus assets and cash flow would not present a problem, but in the period before the Federal Reserve Act any major loan could become a liability and instrument of takeover. Westinghouse ran all his other companies almost debt free, but in the case of Westinghouse Electric, debt was part of doing business. Westinghouse did push the limit with his mammoth projects that maybe in hindsight should have required a more conservative approach. But that's hindsight; the two biggest debt issues were the Chicago Fair and the Niagara project, and these historians would consider them foundational to Westinghouse Electric.

The real disappointment for Westinghouse had to be Pittsburgh itself. Andrew Mellon had exacted his revenge for Westinghouse not bringing him into the operation in the 1880s. Still Westinghouse was a major employer and exporter for the Pittsburgh economy and Mellon should have helped. In fact, after the fall of Westinghouse, Mellon Bank started a purchasing plan of Westinghouse stock to gain a board membership in the 1920s. Henry Clay Frick proved most disappointing having powerful ties to both Morgan and Mellon. The Pittsburgh press attacked as if owned by Morgan. Pittsburgh business associates, like Carnegie and Schwab, spouted the Morgan line of mismanagement. The most hurtful aspect was that the city seemed to believe it with the exception of Wilmerding, Swissvale, East Pittsburgh, and Turtle Creek, who knew better.

The data showing Morgan's manipulation of the Panic of 1907 is more damning, but it is also clear that government needed help to stop the crisis and allowed Morgan to profit in return. Ultimately, the Senate investigations led to the Federal Reserve Act. Many independent financers, such as John Moody checked the panic by "taking a few dollars out of some pockets and putting millions in others." Populist and liberal Wisconsin Senator Robert Lafollette went further, charging "a group of financiers who withhold

108 Arthur Pound and Samuel Moore, *They Told Barron: The Notes of Clarence Barron* (New York: Harper & Brothers, 1930), 83

and dispense prosperity . . . deliberately brought on the late panic, to serve their own ends." In investigations in 1911, Teddy Roosevelt basically said he gave approval to do whatever it took to end the panic. Without a Federal Reserve, he contended that the government would have gone bankrupt. Westinghouse's ownership was therefore both an opportunity for Morgan and a casuality of the Panic.

But is there any truth to the claims of mismanagement? As a business professor, it was an area I was most interested in. Morgan's claims that Westinghouse's companies were too centralized, and the Carnegie concept that Westinghouse was overextended seems to lack proof as a whole. Air Brake, Westinghouse Machine, and Union Switch and Signal functioned as decentralized companies. Some truth to those claims can be found in Westinghouse Electric, which represented most of Westinghouse's interests from 1886 on. Westinghouse Electric required close management of finances and a banking partnership, neither of which interested Westinghouse. He had excelled at bringing in engineers and scientists, allowing them to use their creativity unhindered. He delegated management functions well, but as technically strong as Westinghouse Electric was, it lacked financial expertise. He had developed the corporate concept of a research and development organization, but had antiquated financial management systems. He had beaten Edison's highly centralized Edison Electric and financially controlled General Electric, but his poor attention and inability to bring in the financial expertise allowed for attacks by opportunistic bankers. Westinghouse had strong boards at Air Brake and Westinghouse Machine to keep his project spending in line as well. It is however, too easy to assume Westinghouse's costly projects brought in down. The fact is, his great projects lead to the technology that crushed his competition. Carnegie, Ford, and Rockefeller all disliked bankers, but managed them instead of taking an adversarial position. The bottom line, however, is that the claim of mismanagement cannot stick. His record of forming 60 companies in some twenty countries in very diverse fields is too overwhelming. His business record and accomplishments would only be approached by a handful of people through the ages.

The personal struggle of Westinghouse after his loss of Westinghouse Electric is not well documented, but as we have seen his behavior changed. As a puppet President of Westinghouse Electric, he avoided the labs and factories where he had previously been a daily visitor of. Marguerite played an even larger role in supporting her husband, but this setback ran deep in a man that "had never been defeated." It was a type of depression that few men know. He had come to the heights only to lose what he cherished most in Westinghouse Electric. Close associates, like Benjamin Lamme, reported that Westinghouse, once a daily visitor to East Pittsburgh was rarely seen after the take-over. He was never the same, and the depression seemed to lead to many physical symptoms. Westinghouse had always hated to go to doctors or hospitals but aliments increased after 1907 requiring him to. With Marguerite's help and a real concern for the stockholders (many of which were friends, employees, and family), Westinghouse stayed on as President of Westinghouse Electric and Manufacturing Company. He had to live with the reality that he was no longer in control, and the daily overview bank controlled management . The directors needed to justify their takeover by pointing out mismanagement. To function in such a situation would be difficult for any person. These were difficult years for him, but he tried to involve himself in engineering projects to help ease his depression. Work had always been therapeutic for him.

At the time of the corporate receivership, Westinghouse was immersed in the development of a marine turbine to propel ships. This had been an extension of his boyhood rotary engine. Westinghouse had already brought in two consultants, Admiral George Melville, the retired Chief Engineer of the Navy, and his associate John Macalpine. The idea of a marine turbine went back to Westinghouse's own Navy days when he had experimented with a turbine model boat. Even Parsons had shown some success with marine turbines. Actually, Westinghouse had a steam turbine working at Garrison Alley, since 1894, but it had many problems. The engineers at the time laughing watched these early experiments of Westinghouse in a corner of the building calling the turbine, "The Brass-Chewer" because it tended to "eat" up brass bearings daily.[109] The use of electrical turbine generators on ships had become popular because of their quiet running, low vibration, and low maintenance requirements. Electrical turbo-generators had been installed in the greatest ships of the time, such as the *Lusitania* and *Mauritania.* The use of a marine turbine, however, presented a new set of problems, which was exactly the type of mental tonic Westinghouse needed at this point in his life. The application of a turbine driven propeller required some form of speed reduction. Steam turbines worked naturally at very high speeds. A ship propeller has to work at a much slower speed. High speeds cause a ship's propeller to slip failing to transmit power to move the ship. Also, various operating conditions and different classes of ships required different speeds.

The problem required a mechanism to control and reduce speed. The other problem would be developmental funds, which the new Morgan controlled board watched closely. Typical of the Morganization plan, they were somewhat happy to keep Westinghouse as President but consumed in a project (at least in the first year or so). Westinghouse, however, wanted no interference, and transferred the whole project to Westinghouse Machine Company, in which he exercised full control. It also allowed Westinghouse to slowly distance himself from Westinghouse Electric. Still, Westinghouse Machine Company could no longer use Westinghouse Electric with its huge cash flow to offset capital needs. Within months Westinghouse Machine went down as well, but under a different group of receivers, Westinghouse

109 *Boston Mass Transcript*, March 14, 1914

forged ahead. Henry Herr, Vice-President of Westinghouse Machine, was given the role of project manager. Initial experiments tried to reduce the speed directly, but this also greatly reduced the efficiency of the turbine, which needs high speed. Gearing mechanisms offered the other approach. Macalpine and Melville worked out a gear reducing mechanism that could work with his turbine design. The experimental unit was rated at 3000 horsepower at a cost of $75,000. Experiments showed the unit could develop an amazing 5000 horsepower. By 1909 the Westinghouse team had a 6000 horsepower unit working in East Pittsburgh.

Marine turbine experimental successes led to a contract with the Navy. In 1911 the US Navy 20,000 ton Collier was driven by two twin turbine propellers. It was a great engineering triumph. Money, however, was becoming hard to come by. At East Pittsburgh he built an eighteen-foot deep tank to perform marine engineering experiments. He started a wide array of experiments in areas such an air lubrication to increase the speed of ships. He consulted with Lord Kelvin on the physics involved. While little came out of many of these experiments his biographer, Henry Prout noted For Westinghouse this propeller interlude was a fascinating pastime at a time when he greatly needed diversion, in the darkest moments of his life, when some of his companies were going through receivership." About 1910 Westinghouse started to pursue a completely new idea, which he hoped might lead to a new company in another growth industry.

As noted, Westinghouse had been introduced to the automobile in 1904 with the gift to him of a limousine. Westinghouse was looking at an air spring for the use in an automobile. At the least, cars had no shock absorbers, but depended on spring to cushion the ride. Westinghouse started to design a type of air shock absorber. He took his idea to Henry Herr Mr. Herr, I have a good deal on my mind, but I like to talk to you about these mechanical things. They relieve me."[110] Herr tried to help, but there was little interest to pursue developmentally outside their core business. Westinghouse got some of these devices made at the machine Company, which he installed on his family car at Solitude. Westinghouse was also very interested in electrical cars, which at the time were competitive with gasoline cars. Westinghouse formed a car related company to pursue development in this area. Westinghouse electric motors had been used in early electric car since 1900, but Westinghouse never fully took to the idea of "horseless carriages." Still, in the electric car segment, which was substantial from 1895 to 1905, Westinghouse had established a solid business. Westinghouse continued to involve Westinghouse Electric into electrical needs of the general automobile industry, such as batteries, starters, and small generators. Even after Westinghouse death, this effort continued with Westinghouse Electric buying the Stevens-Duryea Company to manufacture car parts in 1915. Traveling in railroad palace was the ideal of perfect travel for him, yet he saw the

110 Prout, 252

future potential of the gasoline engine. Another idea he was pioneering was the solar engine. He had developed a working solar engine in the desert of Arizonian, and in these later years would visit it to note its operation.

As early as 1909, Westinghouse's health started to deteriorate further. He reported having trouble with climbing the steps to his Air Brake office in Wilmerding. In 1909 Westinghouse Electric Company was free of its financial problems, but Westinghouse was clearly being pushed into an even lesser role. Early biographers note that Westinghouse had "heart problems," but this was a common diagnosis of the time. Probably some combination of depression and health was at play. Westinghouse had tried hard to stimulate the creativity that had carried him, but it was clearly getting harder. Work, for the first time in his life, was stressful. Westinghouse was a creative dynamo that made for a poor employee. Corporate structure inhibited him, where as President, he followed his inspiration. The struggle after 1907 was a different one for Westinghouse, it was political and internal. Westinghouse really headed up a group of corporate insurgents at Westinghouse Electric. The bankers had failed to get Westinghouse to take the six months vacation, they wanted him to. In the end, however, the bankers would need Westinghouse to save the company. Westinghouse represented the stockholders, who were not in line with the bankers. The struggle for a plan of reorganization took 14 months. The initial shock and awe of the banking takeover faded as the stock started to rise in 1908, strengthening the stockholders position. The reorganization required a new stock offering of $6 million dollars for working capital; again, it was Westinghouse that rose to save the company. The bankers and Westinghouse, who represented the stockholders, were at an impasse.

The plan Westinghouse developed was known as the Merchandise Creditors' Plan. Westinghouse like Solomon, understood that the employees did not need the company to fail over politics. Westinghouse, himself, invested $1.5 million of his own money and the employees kicked in $600,000 dollars. The reorganization plan went into effect in December 1908. Westinghouse had brokered a compromise to save the company at his own expense. This, of course, was what the bankers demanded, a limited role for Westinghouse. The new board was more than half controlled by the bankers and creditors, with lawyer Robert Mather (a friend of Morgan) as chairman of the board. The Morgan-Belmont alliance had produced a united banking front, but the banking consortium was a balanced group. Actually, in the end the Pittsburgh bankers would end up in charge, not Morgan. Morgan, however, had achieved his goal of banker control of one of the last Victorian industrial empires. Westinghouse, as noted, moved into a restricted presidency. The bankers failed to get him out, but a proxy fight started to build almost immediately after the reorganization. In January 1910, the board of directors voted Westinghouse a six-month leave of absence, which he again resisted. The internal struggle between Mather and Westinghouse intensified, but Westinghouse was losing ground. Finally, at the July, 1911 Stockholders meeting,

Westinghouse lost the proxy fight, and was ousted as president of Westinghouse Electric. The directors appointed Boston businessman, Edwin F. Atkins, as president. Westinghouse would resign. Ernst Heinrichs reminisces of the fall in 1931 The loss of the Electric Company was to Mr. Westinghouse a disappointment from which he never recovered. There is no doubt that it broke his spirit." [111]

Unfortunately, Westinghouse left with a number of important projects. He had been supplying electric motors for fans and cars, but was also working on a number of futuristic appliances at the time. The coffee maker would be introduced in 1913 along with the electric toaster. He had been working personally on solar energy, which was many years ahead of its time. He had a functioning solar engine in the Arizona desert, which he was following with visits. He had started the company into automotive electrical parts. He was also improving his refrigeration systems. With the exception of the solar engine, he left most of this work to be carried on by Westinghouse Electric.

Westinghouse's health deteriorated rapidly after the 1911 proxy loss. Westinghouse's doctor, William Stewart, requested him to rest, and he cut off most of his business ties by 1912. Westinghouse, who had avoided doctors for most of his life, became close friends with Doctor Stewart. Stewart continued to talk him into longer vacations at Erskine Park. The reported symptoms of sleeplessness, restlessness, and anxiety seemed to support the idea of him suffering from depression. When he was in the office friends reported a quiet man, not the unusual dynamo of past years. The problems compounded in 1912 as Marguerite suffered a stroke of paralysis, which forced the couple to almost full time residence at Erskine. By November of 1913, Westinghouse was free of all duties in Pittsburgh and brought Doctor Stewart to Erskine to live with him. Westinghouse was clearly depressed, but he continued to dream. After the rotary engine of his youth, Westinghouse had pursued the idea of building a machine to abstract heat from the atmosphere. Great scientists of the time called it a violation of the Second Law of Thermodynamics, a perpetual motion machine, and an example of Westinghouse's poor grasp of science. Early biographers point to the many letters of Lord Kelvin telling Westinghouse of the errors in his ways, and pointed it out of as an example of Westinghouse's weakness. Many note that thermodynamics were beyond the understanding of a non-degreed engineer. Westinghouse toyed with this idea to his death. Henry Prout reported that Westinghouse letters reflect he has working on it in 1913.[112] When current engineers review this project we see that Westinghouse was vindicated in the 1980s with the commercial heat pump to warm houses.

In February of 1914, while fishing, Westinghouse fell into the water, which led to a serious cold. For weeks, Doctor Stewart slept in the adjoining chambers, as Westinghouse struggled through each night. He was confined

111 E. H. Heinrichs, "Anecdotes and Reminiscences of George Westinghouse," October 1931, George Westinghouse Museum Archives, Wilmerding, Pennsylvania
112 Prout, 318

to a wheelchair for the period as well. In March arrangements were made for Westinghouse to go to New York to be looked at by specialists. Marguerite made arrangements for a suite of rooms at the Manhattan Hotel Langham over looking Central Park. Actually, Marguerite had started renting a year earlier to be nearer to good doctors for both of them. The plan was to go on to Blaine Mansion if he improved. It was here on March 12, 1914 that he passed peacefully. Marguerite followed him in death in June. The was funeral at Fifth Avenue Presbyterian Church in New York. Floral arrangements came in from kings, princes, and presidents. It was attended by most of Pittsburgh, New York, and Washington's social elites, but eight long time employees carried the casket. Delegates from all the nations' engineering societies were present. George III and his wife would ultimately be the sole heir. George III was a Yale graduate and had married in 1909. The many factories and plants in Europe and America were idled for a few hours to honor their founder. It is estimated that over 100,000 men halted work to honor Westinghouse. In Wilmerding the plant was shut down for two days as many took the train to New York for the funeral. He was buried at Woodlawn Cemetery, but in 1915 both he and Marguerite were moved to a grave in Arlington Cemetery. They had taken many walks together in Arlington while at Blaine House. The simple inscription on the stone was; "Acting Third Assistant, US Navy." The scenes in death reflected the humility of his way of life. The eulogy of pastor Fisher quoted the following:

> "I know the night is near at hand;
> The mist lies low on sea and bay;
> The autumn leaves go drifting by;
> But I have had the day."

At his death he owned or controlled over 15,000 patents, 314 of them his. He had formed over 60 companies, and created millions of jobs worldwide. He had over 50,000 employees working directly for him at his death. He had created the town of Wilmerding and developed many others including East Pittsburgh and Turtle Creek. He had honorary doctorates from Union College and the Königliche Technische Hochschule in Germany. He held the Legion of Honor from France, Order of Leopold of Belgium, and the Royal Crown of Italy. He had won Germany's highest engineering award, the Grashof Medal and America's Edison Medal. Other honors included the John Fritz Medal of the American Association for the Advancement of Science and the Franklin Institute's Scott medal.

EPILOGUE: CAPITALISM WITH A HEART —A FUTURISTIC VISION

> "If someday they say of me that in my work I have contributed something to the welfare and happiness of my fellow men, I shall be satisfied."

The legacy of Westinghouse remains today. He changed industry, technology, and Pittsburgh forever. This was a man loved by his family, employees, and city. More so than Edison, he gave light to the world. His air brake revolutionized the railroads, and his steam engine revolutionized agriculture. He inspired genius, much as Nicola Tesla and Henry Ford. He had won the "war of the currents" making AC current the power source of the nation. He was behind the great Niagara Power Station. In 1930, 55,000 present and former employees erected a memorial on voluntary contributions, fulfilling his own hope — *"If someday they say of me that in my work I have contributed something to the welfare and happiness of my fellow men, I shall be satisfied."*

Westinghouse's view of capitalism is today overlooked, but it may demand a new look in today's globalized market. The plight of the worker during the industrial revolution, eventually, pulled the heartstrings of even the Robber Barons. The solutions took many paths, such as the patriarchical capitalism of Andrew Carnegie, the welfare capitalism of J. P. Morgan, the unionization of Samuel Gompers, and the communal capitalism of Robert Owen. There were even political solutions, such as Communism and Socialism. The Puritan or Christian capitalism of Westinghouse is the most overlooked because of the basis of Big Business. Westinghouse presented innovation, entrepreneurship, and capitalism as linked and interrelated. Like Tesla's Niagara Falls speech of 1897, Westinghouse believed that capitalism would lead to a technical utopia. He had been a strong supporter of William McKinley's "lunch pail republicanism." Capitalism meant that working men could work, and it was work that satisfied men. Westinghouse saw men like Morgan, however, as exploiters of capitalism, not as drivers. The idea of trusts bothered Westinghouse deeply, and he saw trust formation on a par with socialism as a threat to capitalism. Westinghouse's advertising even suggested that customers needed to support Westinghouse Electric against the "electrical trust." In one advertisement Westinghouse Electric asked We invite the cooperation and support of all users of Electrical Apparatus who desire to have the benefits of competition."[113] Westinghouse also believed that government's role was to ensure safety, not be involved in technical decisions. Westinghouse's idealistic view of capitalism was that companies should compete on technology, innovation, and efficiency.

Westinghouse detested legal suits, which were inconsistent with his Puritan roots, yet a lot of his business career was spent in courtrooms. The apparent contradiction arose from his passionate belief in intellectual rights. Respect for intellectual rights was part of the dignity of man as well as the core of his view of capitalism. He had never forgotten his first partners' effort to steal his invention of the railroad frog and car replacer. Without an absolute respect for intellectual rights, capitalism loses its ability to innovate. Certainly, Westinghouse today would be on the forefront of free trade, but he would never deal with a country, like China, that shows no respect for American patent law. He had fought capitalist giants, such as Morgan and Vanderbilt, over this very point. Even more importantly he didn't believe that corporations had rights over individual innovation. To Westinghouse, the individual owned all rights to and rewards from any invention. Bankers stood in shock as Westinghouse paid huge sums for marginal inventions. He never tried to infringe on a patent even when lawyers suggested the possibility. Westinghouse stood tall in a Victorian era, which trampled on individual rights for more profits and industrial "efficiency." Such respect for intellectual rights gained him the loyalty of Nikola Tesla. That bond of mutual respect won him the war of the currents. As noted, the simple plaque at Niagara Falls honors the patents involved.

Few could see that the strength of Westinghouse's organizational research and development was rooted in his respect for individual rights. Westinghouse saw the progress of American technology as the progress of individuals, not corporate research. This Jeffersonian view remains even today the core of our democratic capitalism. Morgan, of course, argued that patent fights wasted resources and money, and stalled technology. Morgan was correct in deducing this from the myriad of electrical patent battles, but he realized that these battles were a result of not respecting individual

113 Westinghouse Electric and Manufacturing Company, Annual Report, May 18, 1892

rights. Unfortunately, the later Victorian view of patents was one of a legal deterrent, which could be circumvented with more improvement. The argument had some validity in the fact that inventions seemed to evolve from streams of innovation, not necessarily a breakthrough event. Furthermore, social Darwinism seemed to imply that the spoils belonged to the one who successfully exploited the innovation. In fact, the crown of invention often went to the commercial developer, such as Bessemer (versus William Kelly), Thomas Edison (versus Joseph Swan), and Nikola Tesla (versus Michael Faraday). The problem is that this socialistic approach to innovation and invention would actually reduce the motivation to innovate in a capitalistic society. Without the economic rewards afforded innovation, it is unlikely that men like Edison, Westinghouse, Thomson, Sprague, and Stanley would have persisted.

Westinghouse battled the Morgan trusts with innovation, technology, and invention. It took the best technology to stay ahead of the trust's purchasing power. The trust was a horizontal and vertical octopus, which fed on itself. Louis Brandeis gave the best description of the Morgan trust in 1911 before Congress:

> J. P. Morgan (or partner), a director of the New York, New Haven, and Hartford railroad, causes that company to sell to J. P. Morgan & Co. an issue of bonds. J. P. Morgan & Co. borrows the money with which to pay the bonds from the Guaranty Trust Co., of which Mr. Morgan (or partner) is a director. J. P. Morgan & Co. sells the bonds to the Penn Mutual Life Insurance Company of which Mr. Morgan (or partner) is director. The New haven spends the proceeds of the bonds in purchasing steel rails from the United States Steel Corporation, of which Mr. Morgan (or partner) is a director. The United States Steel Corporation spends the proceeds of the rails in purchasing electrical supplies from General Electric Company, of which Mr. Morgan (or partner) is a director.

This, of course, overlooks the market efficiencies of trusts, which was Morgan's main justification for trust arrangements. Morgan supporters would argue that this is a worst-case scenario that overlooks the market efficiencies The argument went on for years until 1929 when the Pecora investigation concluded; "the men of Morgan's circle had proved themselves disgracefully devoid of ethics or conscience when it came disposing of the savings of working people."

Westinghouse hunted for inefficiencies on the factory floor versus the boardroom, but Westinghouse understood economies of scale. There were times when he ran projects through parts of his many companies. Westinghouse Electric and Westinghouse Machine Company often acted as a single company, coordinated by Westinghouse himself. Certainly, Westinghouse's patent board agreement seemed a compromise on competitive capitalism. After the patent agreement General Electric and Westinghouse Electric did behave monopolistically in the incandescent light market, which combined the two companies and controlled 95% of the market. Still, Westinghouse believed in competition, but, unlike Morgan, he is saw no need for destructive competition. Their approaches to cooperative advantage were very different, but what stands out is Westinghouse's respect for individuals over corporations. An ethical compass that can only be understood by his Puritan upbringing guided Westinghouse. In 1911, as President of the Society of Mechanical Engineers, George Westinghouse, addressed a conference of American Engineering Societies supporting the anti-trust action of the Taft Administration.

For many years the tendencies have been strongly toward large and powerful railway and industrial combinations. Their very magnitude, coupled with the evil practices so frequently disclosed in press and in our law courts, has so aroused the public that there is now a fixed determination to establish by national and State laws an exacting governmental control of practically all forms of corporations, in order that competition may be encouraged and not stifled, but seemingly with due regard to the real objects in view —the securing of the best public service in all forms, the best foods and goods for our daily needs, the greatest possible comfort to the masses, and as great freedom as possible from those restrictions which hinder rather than promote honest endeavor. Many of the hardships which will arise might had been avoided by those responsible for the creation of great combinations had they appreciated the inevitable consequences of their selfish and unwise course in suppressing competition by the methods transparently wrong. But fortunately there are indications that the great leaders are alive to the importance of the regulation of legislation, businessmen to their senses. The engineering societies, by joint action, have it in their power to do much. Probably there is no better way than to show, from their knowledge and experience, that unregulated competition and rivalry in business have made our costs greater and rendered ideal conditions in industrial and engineering matters most difficult of realization.

This represents one of Mr. Westinghouse's rare public statements on the abuses of trusts that he had resisted throughout his career. He had, as early as 1892, started to call General Electric, the "Electric Trust." He argued in an 1893 advertisement We invite the cooperation and support of all users of electrical apparatus who desire to have the benefits of competition."

Yet, Westinghouse understood well the inefficiencies of competition in the marketplace. Patent battles had been costly to all in the electrical industry. These legal battles, not only cost huge sums, but slowed progress or wasted time in legal solutions. The waste of legal costs was something that both Morgan and Westinghouse agreed on. Of course, Morgan's solution was control by non-competitive trusts. Westinghouse called for a bigger role of the engineering societies in limiting these battles and settling them in the technical arena. Westinghouse had seen the great outcry of engineers when Professor Forbes blatantly was given the generator contracts by impinging on Westinghouse's patents. This visionary role of the engineering societies has never really come to fruition. However, Westinghouse did correctly foresee the role of engineering societies in technical product standardization.

One of the issues that Westinghouse was well aware of was that competing technologies on the same product often result in non-standardized parts. The cost of replacement parts therefore restricted competition in the market place. Companies were forced to go with one system and once there, they had a high conversion cost to try a competing system. Westinghouse put it best in his own words No user of electrical apparatus can fail to appreciate the advantage it would be to him, when some repair part is needed, if certain standards were followed by all constructors with reference to equivalent devices; but it is lamentable to say that with the simple exception of uniform bases for incandescent lamps, there are now practically no standards [1910]. The vast majority of our inventors proceed along independent lines, with the result of a constantly growing confusion, to the disadvantage of everybody." This type of market inefficiency can be reduced by standard organizations. Still, it is used for competitive advantage in emerging markets. Apple Computer and Sony beta both used a lack of standards to maintain a commercial edge, but ultimately the market will force standardization.

Westinghouse's view of capitalism could be seen in his somewhat inconsistent politics. Certainly, for most of his life he was a Republican, but he had crossed over for democratic Grover Cleveland. Cleveland's pro-business approach had even got the support of the Pittsburgh local Republican Party. Westinghouse's Republican roots went back to abolition and Lincoln. Cleveland had offered an anti-boss, anti-trust approach that had won over many Republicans in 1892. Cleveland believed also in a full employment approach, it was a popular movement that the Republicans would later on adopt as their own. Westinghouse's close friend in the Senate had been Republican William McKinley. McKinley came from industrial Canton, Ohio, which sun dogged Pittsburgh. Canton was an industrial town with many railroads. McKinley had worked with Westinghouse on railroad and industry safety. They both favored a soft government hand to keep business ethically straight. McKinley was a "lunch pail" Republican. McKinley wanted full employment, and was willingly to take whatever measures needed. Work for both men was the opium of the masses. This blend of job based polices, like for Democrat Cleveland with the Republicans, caused much cross over among the Democrats to McKinley. In a lot of ways the McKinley Democrats were reassembled as the Reagan Democrats of the 1980s. He, like Westinghouse, favored tariffs only when proven to be needed to foster job creation. Both McKinley and Westinghouse opposed the trusts and unethical behavior, while being pro-business. Both believed in a stronger application of the 1890 Sherman Anti-trust Act that had been nullified by the courts. Both men felt unions were to be opposed, but more as a reaction to poor management. Both were always anti-union but pro-worker. McKinley believed in tolerance among religions while being a somewhat independent believer like Westinghouse. McKinley was also one of the first supporters of women rights. McKinley, however, like his predecessors and followers, kissed the ring of Morgan. Morgan was just too dominant to be opposed by either Republican or Democrat.

Westinghouse referred to his employees as "fellow-workers," yet he was no socialist or communist, but a capitalist in every sense of the word. He was compassionate and believed in full employment, but he believed that owners deserved part of the spoils of their effort. Capitalism was the great motivator and strength of democracy. He, of course, did not believe that financiers deserved the amounts they were taking in during the Gilded Age. Bankers were to Westinghouse unnecessary middle men. Part of Westinghouse's lunch pail Republicanism was an anti-union plank. This may seem strange to the modern reader that such a pro-worker employer would oppose unionism. Unions represented a third link in the corporation that Westinghouse saw as unnecessary, and under Westinghouse's helm so did his employees. We have seen that Samuel Gompers had praised Westinghouse's approach. Westinghouse was not, however, outspoken as an industrialist on the subject. He realized by the views of Braddock and Homestead from his daily ride on the Pennsylvania Railroad that many exploited the worker. Unions sometimes were a necessary evil. Westinghouse wanted to avoid the blacklisting, yellow dog contracts, and the Pinkerton guards of the neighboring factories of Braddock, McKeesport, and Homestead. In this respect, neighboring Wilmerding was more like a different country.

J. P. Morgan had a low opinion of most of America's founding inventors and capitalists, such as Edison, Carnegie, Ford, Rockefeller, and Westinghouse. These men viewed physical assets as the basis for wealth, while Morgan saw financial capital as playing the main role. Morgan seemed to take pleasure in taking these empires down and putting in his own management. Westinghouse represented the most difficult for bankers such as Morgan to control because of the passion, love, and ethical commitment to his great works. In 1900 Carnegie was one of Morgan's biggest trophies after Vanderbilt's railroad empire. Officially, Morgan bought out Carnegie at a price of Carnegie's setting, but Frick claimed Morgan's liberal price was for the "elimination of Mr. Carnegie." Morgan had often claimed that Carnegie had "demoralized the steel industry." Henry Clay Frick, as a Morgan associate and enemy of Carnegie, claimed victory openly.[114] Frick had worked behind the scenes to help Morgan enter the steel business in the 1890s. Morgan's disrespect for Carnegie's management ability might be more of Morgan's publicity campaign to gain public support for the United States Steel merger. Morgan always tried hard to appear as the guardian angel of America's industry. Frick's hatred of Carnegie was clear, and as a director of United States Steel, he helped to purge the company of "Carnegie men." In 1901 Frick built the 22 story "Frick Building" beside the 14-story Carnegie Building in downtown Pittsburgh, so as to cast a shadow on the Carnegie Building. In New York Frick built his mansion down from Carnegie's with the purpose

114 Martha Frick Symington Sanger, Henry Clay Frick: An Intimate Portrait (New York: Abbeville Press Publishers, 1998), 291

of "making Carnegie's place look like a miner's shack." A lot of Morgan's dislike for Carnegie's management style came from Frick's new position as a director of USS. Carnegie, busy improving his own image by giving away his money, said little of his true feelings about Morgan.

Westinghouse's view of bankers was well known, but it never appeared in print. These views can best be summarized by Henry Ford's published views:

> And that is the danger in having bankers in business. They think solely in terms of money. They think of a factory of making money, not goods. They want to watch the money, not the efficiency of production . . . Bankers play far too great a part in the conduct of industry. Most business will privately admit that fact. They will seldom publicly admit because they are afraid of their bankers. It required less skill to make a fortune dealing in money than dealing in production. The average successful banker is by no means so intelligent and resourceful a man as is the average businessman. Yet the banker through his control of credit practically controls the average business man. . . The banker is, as I have noted, by training and because of his position, totally unsuited to the conduct of industry.[115]

Westinghouse might have added that bankers arrived at their position by birth versus effort. Westinghouse was never afraid of bankers.

Interestingly, McKinley, Morgan, and Westinghouse all struggled with and in the Republican Party. The Sherman Anti-Trust Act of 1890 exemplified the paradox of these men. Men like Westinghouse and McKinley opposed trusts, but on the other hand, the Sherman Act seemed to violate the Protestant work ethic. All of them believed the capitalist should be rewarded for his efforts, and were pro-business. Originally, the Clayton Act was utilitarian, to be applied only in extreme cases of monopolies affecting the national economy. For years the application of the Sherman Act was avoided. Westinghouse favored ethical internal constraints versus legal restrictions. Even Teddy Roosevelt, McKinley's vice-president, talked more trust busting than actual busting. Westinghouse did not oppose bigness as many did; he opposed unfair practices. Westinghouse had, in late 1892, asked his lawyers to argue the Sherman Anti-Trust Act against the "most vicious trust" of General Electric. This occurred after General Electric's Edison-style bulb patent victory. After that General Electric did not ask Westinghouse for royalties, but moved to shut down availability to Edison-style bulbs in an effort to hurt and destroy Westinghouse's Fair project.

Westinghouse demanded respect for his patents, but never was vindictive in his control of his patents. He was tough, but fair. He would have been happy, for example, to pay General Electric whatever they asked to produce Edison-style bulbs for the World's Fair. He rarely wasted his research resources in finding technical loopholes in patents, but preferred to break new ground. If he felt, he might use an other man's patent, he paid the owner fairly, if not liberally. In the final analysis, Westinghouse was neither a politician nor a crusader. Even with the loss of Westinghouse Electric, Westinghouse

115 Henry Ford, *My Life and Work* (North Stratford: Ayer Company, 1922), 176

favored quiet retreat versus public battle. This was similar to McKinley's approach, which is why they became friends. Even Morgan and Westinghouse shared the same politics all their lives. Morgan felt that in the business arena, any type of competition was ethical, within reason. If, through patent control, you could break a competitor, this was fair competition. The ultimate goal of competition was to eliminate competition through a controlled trust. Westinghouse resented the role of financiers. Stockholders were the owners, in his view, not the debt holders. Westinghouse gauged his success not in his profits but in the size of his payrolls. In this sense, like Henry Ford after him, he was a pure industrialist.

Westinghouse was a patriotic American, proud of his Navy service, but he had a global view of the market and respected international competition. Westinghouse generally did not support tariffs, since much of his business was overseas. While Westinghouse manufactured the world over, he favored American production whenever possible. He also favored putting American engineers and managers in charge of his foreign operations. This turned out to have been a major mistake, and Westinghouse had to reorganize in the 1900s to put more local managers in charge. Westinghouse dominated his European competitors with innovation, process efficiency and product efficiency. Still, he would import parts where needed to stay competitive.

Westinghouse preferred not give food to the poor, but tools and jobs. He didn't believe that as a capitalist he was a steward of funds, as Carnegie did. Almost all of Westinghouse's wealth was in his companies, and similarly it was through his factories that he distributed it. He helped his employees with house ownership, pension, health expenses, and social needs. He gave directly to education, but even there, he preferred to donate equipment and generators to schools and educational institutions. He avoided the direct cash payment where possible, believing it was better to pay his employees well and create jobs. Carnegie, on the other hand, reaped huge profits at low wages, and then redistributed to society. There was a certain arrogance in Carnegie's approach, and a certain idealism in Westinghouse's. It may explain the paradox that Carnegie was loved by the community and hated by his employees, while Westinghouse was loved by his employees and unknown to the general community of the Pittsburgh area.

It would, however, be a mistake to suggest that he did not give liberally to the less fortunate. His giving was generous to individuals, but always undocumented, by his own request. Mrs. Westinghouse followed the same credo, but gave liberally as well. They gave quietly, if not anonymously. His friend and partner reported on the other unknown part of Westinghouse's charity:

> It is a matter of history, of course, how Mr. Westinghouse carried out this idea. Thereafter his apparent ambition to build up large concerns had a different aspect in my eyes, as I understood the ethical impulse underlying it. While he disclaimed belief in the efficacy of benevolent giving, and shrank from acknowledgement of his kindness, those of us who were close-

ly connected with him knew of many instances where he was supporting whole families and doing other deeds of helpfulness in an unostentatious way. Mrs. Westinghouse was very sympathetic and loved to relieve distress, and Mr. Westinghouse made her a regular allowance for the gratification of her desires in this respect. The amount was stated to me, and it was large.[116]

116 Leupp, 248

Bibliography

Adams, E. D., *Niagara Power* (New York, 1927)

Alexander, William, *A brief History of the Equitable Society* (New York: Equitable Life, 1929)

American Society of Mechanical Engineers, *George Westinghouse Commemoration* (New York: 1937)

Abraham, J. and R. Savin, *Elihu Thomson Correspondence* (New York: Academic Press, 1971)

Baker, E., *Sir William Preece: Victorian Engineer Extraordinary* (London: Hutchinson, 1976)

Baldwin, Leland, *Pittsburgh: The Story of a City* (Pittsburgh: University of Pittsburgh Press, 1937)

Beard, Patricia, *After The Ball* (New York: Perennial, 2003)

Behrend, B., *The Induction Motor* (New York: McGraw Hill, 1921)

Bridge, James, *The Inside Story of the Carnegie Steel Company* (New York: Aldine Books, 1903)

Bright, A., *The Electric Lamp Industry* (New York: Macmillan, 1949)

Bryan, G. S., *Edison: The Man and His Work* (New York: 1926)

Chernow, Ron, *The House of Morgan* (New York: Atlantic Monthly Press, 1990)

Conot, R., *Streak of Luck: The Life Story of Edison* (New York: Bantam Books, 1980)

Clark, T., *The American Railway* (New York: Charles Scribner's Sons, 1897)

Cowles, Virginia, *Astors* (New York: Knopf, 1979)

Crane, Frank, *George Westinghouse: His Life and Achievements* (New York: Wise & Co., 1925)

Davison, Mary, *Annals of Wilkinsburg* (Wilkinsburg: The Group, 1940)

Eggert, Gerald, *Steelmasters and Reform* (Pittsburgh: University of Pittsburgh Press, 1981)

Evans, Henry Oliver, *Iron Pioneer: Henry Oliver* (New York: E. P. Dutton & Co., 1942)

Davis, John, *The Guggenheims: American Epic* (New York: Morrow, 1978)

De Borchgrave, Alexandra Villard, and John Cullen, *Villard: The Life and Times of an American Titan* (New York: Doubleday, 2001)

Derry, T. and Trevor Williams, *A Short History of Technology* (New York: Oxford Press, 1960)

Dickson W. and A. Dickson, *The Life & Inventions of T. A. Edison* (New York: Crowel, 1892)

Dunlop, M., *Glided City* (New York: William Morrow, 2000)

Hammond, J., *Men and Volts: The Story of General Electric* (New York: J.B. Lippineott Company, 1941)

Harvey, George, *Henry Clay Frick: The Man* (Privately Printed, 1936)

Haskins, C., *General Electric Meters* (General Electric, 1897)

Hawkings, L., *William Stanley, His Life and Times* (New York: Newcomen Society, 1939)

Hesson, Robert, *Steel Titan: The Life of Charles Schwab* (New York: Oxford Press, 1975)

Higgs, Paget, *The Electric Light* (London, 1879)

Howell, J. and Henry Schroeder, *History of the Incandescent Lamp* (New York: 1927)

Hughes, T., *Networks of Power* (Baltimore: Johns Hopkins University Press, 1983)

Jaffe, Bernard, *Men of Science in America* (New York: Simon and Shuster, 1946)

Jonnes, Jill, *Empires of Light* (New York: Random House, 2003)

Josephson, Matthew, *The Robber Barrons* (San Diego: Harcourt, Brace & Co., 1962)

Kirkland, Edward, *A History of American Economic Life* (New York: F. S. Crofts, 1934)

Klein, Maury, *The Life and Legend of Jay Gould* (Baltimore: Johns Hopkins University Press, 1986)

Kuhn, T., *The Structure of Scientific Revolutions* (Chicago: University of Chicago Press, 1970)

Lamme, George, *Unwritten History of Braddock's Field* (Pittsburgh: Nicholson Printing, 1917)

Lamme, B. G., *Electrical Engineer: An Autobiography* (New York: G. Putman's Sons, 1926)

Lamme, B. G., *Electrical Engineering Papers* (East Pittsburgh: 1919)

Leonard, J., *Loki: The Life of Charles Proteus Steinmetz* (Garden City: Doubleday, 1928)

Levine, I.E., *Inventive Wizard: George Westinghouse* (New York: Julian Messner, 1962)

Leupp, Francis, *George Westinghouse: His Life and Achievements* (Boston: Little, Brown, and Company, 1918)

MacLaren, Malcolm, *The Rise of the Electrical Industry during the Nineteenth Century* (Princeton: Princeton Press, 1943)

Mason, E., *The Street Railway in Massachusetts* (Cambridge: 1932)

Maxim, Hiram, *My Life* (London: 1915)

Morris, Charles, *The Tycoons* (New York: Henry Holt and Company, 2005)

Morris, Edmund, *Theodore Rex* (New York: Random House, 2001)

Mottelay, P., *The Life and Work of Hiram Maxim* (New York, 1920)

Newton, John, *A Century and a Half of Pittsburgh and Her People* (Pittsburgh: Lewis Publishing, 1908)

Lorant, Stefan, *Pittsburgh: The Story of an American City* (Lenox, Massachusetts: Authors Edition, 1975)

Passer, Harold, *The Electrical Manufacturers* (New York: Arno Press, 1972)

Prout, Henry, *A Life of George Westinghouse* (New York: Cosmo Classics, 1921)

Reeves, W., *The First Elevated Railroads in Manhattan and the Bronx* (New York: New York Historical Society, 1936)

Rousmaniere, John, *The Life and Times of the Equitable* (The Stinehour Press, 1995)

Sanger, Martha Frick Symington, *Henry Clay Frick* (New York: Abbeville Press, 1988)

Satterlee, H., *J. Pierpont Morgan* (New York: Macmillan, 1939)

Seifer, Marc, *Wizard: The Life and Times of Nikola Tesla* (New York: Citadel Press, 1998)

Schreiner, Samuel, *Henry Clay Frick: Gospel of Greed* (New York: St. Martins Press, 1995)

Smith, Page, *The Rise of Industrial America* (New York: McGraw-Hill, 1984)

Sprague, Harriet, *Frank J. Sprague and the Edison Myth* (New York: William Fredrick Press, 1947)

Standiford, S., *Meet You in Hell* (New York: Crown Publishing, 2005)

Steinmtez, C., *General Lectures on Electrical Engineering* (Schenectady, New York, 1908)

Strouse, Jean, *Morgan: American Financier* (New York: Perennial, 2005)

Usher, A., *A History of Mechanical Inventions* (New York: 1929)

Walker, J., *Fifty Years of Rapid Transit* (New York: 1918)

Warren, Kenneth, *Triumphant Capitalism* (Pittsburgh: University of Pittsburgh Press, 1996)

Woodbury, David, *Beloved Scientist: Elihu Thomson* (New York: McGraw-Hill, 1944)

Archival Records

Edison Papers, Rutgers University

Making of America, Cornell University Library

Historic Pittsburgh, Heinz Regional History Center

Edison Works, Greenfield Village and The Henry Ford

George Westinghouse Museum, Wilmerding

Index

C

D

G

H

I

J

K

L

M

N

O

P

Pittsburgh, 1-6, 7-9, 13-16, 18-23, 28, 37-39, 41, 43-45, 47-49, 53, 55, 57-58, 60, 62, 65-66, 68-75, 78, 80, 84-86, 94, 97-99, 101-102, 104, 107-109, 115-119, 122-123, 127-131, 133-134, 136, 138-139, 141-143, 146, 148, 150-153, 157-159, 165, 169-170, 174-176, 177-183, 187-188, 189, 191-193, 196-197, 199-201, 203-207, 209-210, 213, 215, 217-220, 224, 226, 227, 229-231, 233-236, 237, 241-242, 244, 247-250

Schenley Park, 2-3, 5, 224

T

U